"This book is a deeply researched response to attacks on free will by some extremely dogmatic physicists. After an extensive well-informed survey of relevant issues, the author makes a convincing case that free will is indeed compatible with present-day physics. Highly recommended."

George F.R. Ellis, *University of Cape Town, South Africa*

"Jeffrey Koperski's *Physics and Free Will* is clear, engaging, and offers a compelling argument that an ontological account of emergence is the way to understand how free will exists in a physical world."

Robert C. Bishop, *Wheaton College, USA*

"An original and accessible account of how free will can be made compatible with scientifically minded skepticism, including that of down-to-earth physicists."

Jan Scheffel, *KTH Royal Institute of Technology, Sweden*

Physics and Free Will

Can libertarian free will survive our best physical theories? While prominent physicists argue that genuine agency contradicts science, this book demonstrates that contemporary physics actually provides robust foundations for human freedom.

Physics and Free Will rigorously navigates the complex intersection of physical theory and the free will debate. Koperski begins by clarifying determinism's scientific foundations before systematically addressing challenges from special relativity and refuting physics-based arguments against mind-body dualism. Drawing on insights from continuum mechanics and condensed matter physics, he reveals how multiscale systems exhibit the emergence of causally efficacious properties irreducible to their micro-constituents. Using these insights, Koperski develops a framework that preserves libertarian intuitions about genuine choice while remaining consistent with physical theory. Through meticulous examination of determinism, reductionism, and the causal closure of physics, this comprehensive work reveals how contemporary science's recognition of emergent causation provides ample space for genuine human agency.

This book will be essential for researchers and graduate students in metaphysics, philosophy of physics, philosophy of science, and philosophy of mind, and for anyone seeking to understand how contemporary science reshapes rather than eliminates the possibility of free will.

Jeffrey Koperski is Professor of Philosophy at Saginaw Valley State University. His two previous books are *The Physics of Theism:* God, Physics, *and the Philosophy of Science* (2015) and *Divine Action, Determinism, and the Laws of Nature* (Routledge, 2020).

Routledge Studies in the Philosophy of Science

Conjunctive Explanations
Nature, Epistemology, and Psychology of Explanatory Multiplicity
Edited by Jonah N. Schupbach and David H. Glass

The Aesthetics of Scientific Experiments
Edited by Milena Ivanova and Alice Murphy

Scientific Explanation, Causality, and Agency
A Free Energy Account
Majid D. Beni

Understanding and Conscious Experience
Philosophical and Scientific Perspectives
Edited by Andrei Ionuț Mărășoiu and Mircea Dumitru

Modeling the Possible
Perspectives from Philosophy of Science
Edited by Tarja Knuuttila, Till Grüne-Yanoff, Rami Koskinen and Ylwa Sjölin Wirling

Pragmatism and Philosophy of Science
Historical and Contemporary Perspectives
Edited by Roberto Gronda, Marianne Janack, and Giancarlo Marchetti

Physics and Free Will
The Science of Libertarian Freedom
Jeffrey Koperski

For more information about this series, please visit: https://www.routledge.com/Routledge-Studies-in-the-Philosophy-of-Science/book-series/POS

Physics and Free Will

The Science of Libertarian Freedom

Jeffrey Koperski

Routledge
Taylor & Francis Group
NEW YORK AND LONDON

First published 2026
by Routledge
605 Third Avenue, New York, NY 10158

and by Routledge
4 Park Square, Milton Park, Abingdon, Oxon, OX14 4RN

Routledge is an imprint of the Taylor & Francis Group, an informa business

ISBN: 978-1-041-15791-5 (hbk)
ISBN: 978-1-041-15790-8 (pbk)
ISBN: 978-1-003-68119-9 (ebk)

DOI: 10.4324/9781003681199

Typeset in Sabon
by Newgen Publishing UK

To J.P. Moreland and Philip West, for throwing some light on the path and sending me on my way.

Contents

Preface

"I will not discuss libertarianism because, well, it's incompatible with what we know about nature" (Hossenfelder 2022, 128). This book refutes that claim.

"Libertarianism" is the technical name for the intuitive sort of free will that most people assume they have until philosophical or scientific objections are raised. Whatever action you choose at a given point in time, libertarians typically believe that you could have made a different choice even if the circumstances were the same. Our decisions are not merely the inevitable result of prior causes. Several public-facing physicists insist that libertarian freedom is incompatible with the laws of nature. They are wrong, and I aim to show why.

That's rather bold, isn't it? Why would a philosopher challenge a group of well-known physicists? The answer is partly biographical. As an undergraduate, philosophy didn't hold much interest for me. In fact, the only compelling topic in the one course I took was free will and determinism. It *seems* like we have free will, and the objections presented at that point did not convince me otherwise. Graduate school was another matter. There arguments for why we are not as free as it appears lurked around every corner. In philosophy of religion, the Abrahamic faiths typically say that God infallibly knows the future. No decisions made today can change the future, otherwise God's current beliefs would be false. In philosophy of mind, there is neuroscientific evidence that our decisions do not cause bodily motion. Both action and the sense of deciding to act are effects produced by the brain. In metaphysics, a popular view of time entails that the future fully exists and cannot be changed by decisions made in the present. And there are several arguments from philosophy of physics, which takes us back to the Sabine Hossenfelder quote I started with. Am I free to finish this paragraph rather than getting another cup of coffee? There are many reasons to believe I am not. Philosophers of science are well suited to address this last set of objections because, as we will see, those objections are mainly philosophical, even when they are presented by scientists.

Criticisms will be raised throughout by fundamentalists, but not the sort of fundamentalists you probably have in mind. Fundamentalism in this context is the view that only certain types of research are about fundamental matters. The term often involves a kind of signaling. "Those folks over there do applied stuff, which is fine, but my research is in *fundamental* physics." The fundamentalist believes that all real physics is bottom up and that all phenomena above the fundamental level are reducible to those fundamental truths. Physicist and historian Clifford Truesdell described this mindset years ago:

> Ignorance commonly vents itself in expressions of contempt. Thus "physics", by definition, is become [*sic*] exclusively the study of the structure of matter, while anyone who considers physical phenomena on a supermolecular scale is kicked aside as not being a "real" physicist. Often "real" physicists let it be known that all gross phenomena easily could be described and predicted perfectly well by structural theories.
>
> (Batterman 2021, 9)

This fundamentalist perspective pervades much of contemporary physics, shaping how physicists approach questions about consciousness, agency, and free will. Many are so steeped in this perspective that they have difficulty entertaining contrary points of view. It's just so obviously true that anyone raising objections must either not understand the science or have some other agenda.

Fundamentalism is one of the main targets of this book as it leaves no room for free will. But, I will argue, fundamentalism is either false or at least not rationally required once one understands the relevant physics. In the end, science provides plenty of space for libertarian freedom. It's questionable philosophy that is doing all the work against free will.

That, in short, is the task at hand. The next step is to see why some physicists believe libertarian free will is impossible.

Before jumping in, let me make some acknowledgments. Jeffrey Chalmers, Taylor Cyr, Al Lent, J. Brian Pitts, and Philip Ratzsch all provided useful feedback on chapter drafts. Special thanks to Robert Bishop, Kelly James Clark, Del Ratzsch, and Philip West for commenting on the entire manuscript. Robert made sure I got the technical details right. Del reined me in when rhetorical exuberance got the best of me. Kelly and Phil added clarity throughout. Their work is deeply appreciated.

Works Cited

Batterman, Robert W. 2021. *The Middle Way: A Non-Fundamental Approach to Many-Body Physics*. Oxford: Oxford University Press.

Hossenfelder, Sabine. 2022. *Existential Physics: A Scientist's Guide to Life's Biggest Questions*. New York: Viking.

1 Physics and the Problem of Free Will

1.1 Introduction

While this book is, of course, a fascinating read, you will nonetheless be tempted to check your email or social media feed before the end of the chapter. Are you free to do so? If you choose to continue to the end, could you have done otherwise? Several well-known physicists say "no," there is no room in physical reality for that sort of free will. As Sean Carroll puts it,

> There's no reason to accept libertarian freedom as part of the real world. There is no direct evidence for it, and it violates everything we know about the laws of nature. In order for libertarian freedom to exist, it would have to be possible for human beings to overcome the laws of physics just by thinking.
>
> (Carroll 2016, 381)

That seems bad. If free will requires breaking a law of nature, we probably don't have it.

Carroll mentions libertarian freedom, which has nothing to do with libertarianism as a political philosophy. So what is it? I often tell students it's the sort of free will they think they have until some philosopher or scientist tries to talk them out of it. Libertarian freedom is fully caffeinated free will. It says that our actions are caused in part by our own choices. I could keep writing this paragraph or stop to get some coffee. In some critical sense, which I choose is up to me. Even if I chose to get some coffee, I didn't *have* to do that. I could have done otherwise. Libertarians also deny that the choice was determined either by my brain chemistry or prior psychological states. My free choice was not merely the laws of nature working themselves out through this particular biological system (my body). Libertarians believe that a conscious human person is an agent, and that agent's choices are to some degree within their control. Such freedom,

DOI: 10.4324/9781003681199-1

however, is not unlimited. We are not free to levitate, for example, or hold our breath for more than fifty minutes. But within such limits, the future is open, and how it plays out depends in part on our choices. Chapter 2 includes other varieties of free will, but libertarianism is the most intuitive.

Why then do many physicists, especially public-facing physicists like Carroll, deny libertarian freedom? Answering that question is the goal of this chapter. Let's begin with a bit of history.

1.2 Historical Roots

Those who deny free will have historically affirmed determinism. If determinism is true, then there is only one way that future events can go, including one's desires and choices. Libertarian freedom would be an illusion in such a world, akin to how some philosophers think of color or sound—so-called "secondary properties" that are only in our experience. We say the explosion is loud, but the sound itself is merely a matter of one's internal phenomenology. It is not "out there" in the world, unlike the compression waves that produce it. Likewise, perhaps free will is just a feeling.

Ancient Stoicism took steps in this direction. Rather than determinism per se, the Stoic view is rooted in providence (Gr. *pronoia*). They believed in a pantheistic god who completely orders events, leaving no possibility that humans might bring about a different outcome (O'Keefe 2021, sec. 1.4). If God were to ordain for me to become a philosophy professor in the twenty-first century, I would inevitably do so. The Stoics claimed that there was still some measure of individual freedom in their "choices," however. So long as a person's choices coincide with what God has determined, that person is free to act accordingly. When we choose otherwise, God ordains different circumstances to bring about the same end. If I were not to apply to graduate school, for example, God would close off other means for me to make a living until I choose to make the choices that coincide with God's will.

Stoicism influenced the theological doctrine of providence in early Christianity. In contrast to pantheism, the Christian God is an omnipotent and omniscient deity who could exercise a fine-grained sense of providential control. While the earliest Church Fathers rejected Stoic determinism, Augustine came to fully lean into it (Fergusson 2019, 49–51):

> But, as to those who call by the name of fate, not the disposition of the stars as it may exist when any creature is conceived, or born, or commences its existence, but the whole connection and train of causes which makes everything become what it does become, there is no need that I should labor and strive with them in a merely verbal controversy, since they attribute the so-called order and connection of causes

> to the will and power of God most high, who is most rightly and most truly believed to know all things before they come to pass, and to leave nothing unordained; from whom are all powers, although the wills of all are not from Him. Now, that it is chiefly the will of God most high, whose power extends itself irresistibly through all things which they call fate...
>
> (Augustine 1887, vol. 2, bk. 5.8)

He then quotes the Stoic Seneca with approval. Augustine takes the Stoic view, often called "fate," to be correct up to a point. God leaves no event "unordained." Augustine's views were in turn highly influential on the Protestant Reformers (Fergusson 2019, sec. 2.3).

Today, all of the Abrahamic faiths affirm a doctrine of providential guidance, but they differ in the degree to which God determines events.[1] Some hold a wide sort of providence whereby God has set up general laws to produce food and resources for our use. Others believe in meticulous providence in which literally everything that occurs is ordained by God. Surprisingly perhaps, all sides also typically affirm free will. According to The Westminster Confession,

> God the great Creator of all things doth uphold, direct, dispose, and govern all creatures, actions, and things, from the greatest even to the least, by His most wise and holy providence, according to his infallible foreknowledge, and the free and immutable counsel of His own will, to the praise of the glory of His wisdom, power, justice, goodness, and mercy.
>
> (5.1)

But also, "God hath endued the will of man with that natural liberty, that it is neither forced, nor, by any absolute necessity of nature, determined to good, or evil" (9.1). Whether these dual affirmations are logically consistent is a matter of debate (Hasker 2004).

By the seventeenth century, the question of determinism moved from theology to natural philosophy. Having rejected the organic, Aristotelian view of nature, the early moderns replaced it with a mechanistic universe. Physical reality, they thought, is machine-like, where pushing or pulling on one part of a system dictates effects elsewhere, as when the gears in a clock determine the position of its hands. Such a view of nature implied causal determinism: every effect has a cause, and nothing happens purely as a matter of chance. Many philosophers wrestled with how or whether free will had a place in a mechanistic universe.[2]

It took some time before mechanistic philosophy could be refined into mathematical physics. While Descartes proposed three laws of motion decades before Newton, he did not believe that mathematics could describe the whole of nature (*Principles of Philosophy* III, §57–61). Only particularly simple systems could be accurately described using equations. This pessimism carried through at least as far as D'Alembert in the eighteenth century (Truesdell 1984, 80–83; Wilson 2000, 298–301). So, while mechanistic philosophy was widely accepted, it wasn't yet clear whether physics could prove the universality of causal determinism. In time, Pierre-Simon Laplace argued that physics reveals nature to be deterministic and that classical mechanics could, in principle, predict the universe's entire future evolution (Laplace [1814] 1902, 3–4). Although Laplace himself did not explicitly say that determinism in physics negated free will, many have taken that to be an implication—an issue we will take up again in Chapter 3. Suffice for now to say that determinism in physics is not new, and with it comes skepticism about free will, at least of the libertarian variety.

1.3 Physics and Free-Will Skepticism

Let's now move into more recent arguments. They are all instances of what Daniel Dennett says he has learned from many scientists: "[Free] will, in their view, is obviously incompatible with naturalism, with determinism, and very likely incoherent against any background, so they cheerfully insist that of course they don't have free will" (Timpe and Jacobs 2015, 319).

1.3.1 Brian Greene

For many physicists, the central issue is less about causal determinism and more about the scope of the laws of nature. No physical events, they claim, lie outside the control of nature's laws. If humans are wholly material beings, as biology and neuroscience assume, then the behavior of the particles that make us up is determined by those laws:

> Some … are then bewildered by the conflict between the freedom of will they experience and the unyielding control of physical law they recognize. But the fact is there is no conflict because there is no transcending of physical law. There can't be. Instead, the collections of particles need to reassess their powers, focusing not on the laws that govern particles themselves but on the high-level, thoroughly complex, and extraordinarily rich behaviors each collection of particles—each individual—can exhibit and experience. And with that reorientation, the particle collections can tell an illuminating story of wondrous behaviors and experiences, suffused with wills that feel free and speak as though they

> have autonomous control, and yet are fully governed by the laws of physics.
>
> (Greene 2020, 150)

Greene does not present a defense for the thesis that all events are law-governed or for the belief that we are nothing but collections of particles. In any case, these are commonly held views within the scientific community. Given the scope of the laws of nature, free will is at best "an illuminating story." While it is presumably fine to "speak as though [we] have autonomous control" (i.e., no harm done), it is not literally true. Given the way the laws work, it can't be.

Sabine Hossenfelder agrees with the notion of law-governedness but puts greater emphasis on the principle of reductionism.

1.3.2 Sabine Hossenfelder

Other like-minded physicists go a bit further. Not only is a person nothing but a collection of particles, says Hossenfelder, but also the behavior of the whole is dictated by its parts. That this is the case is no longer up for debate, she says, if one knows the relevant science:

> A lot of people seem to think it is merely a philosophical stance that the behavior of a composite object (for example, you) is determined by the behavior of its constituents—that is, subatomic particles. They call it reductionism or *materialism* or, sometimes, *physicalism*, as if giving it a name that ends in *-ism* will somehow make it disappear. But reductionism—according to which the behavior of an object can be deduced from ("reduced to," as the philosophers would say) the properties, behavior, and interactions of the object's constituents—is not a philosophy. It's one of the best-established facts about nature.
>
> (Hossenfelder 2022, 82–83)

Hossenfelder misses some subtle differences between the terms here. Materialism is an ontological position: the only things that exist are made up of material stuff, typically atoms. That was a common view in the age of classical mechanics.[3] Physicists later began talking about fundamental reality in more exotic terms, including bosons, fermions, and a menagerie of particles and force carriers in the Standard Model. Many physicists would now say that even entities that appear to be particles are really just manifestations of quantum fields.

Instead of "materialism," philosophers began using "physicalism": the only things that exist are made up of the stuff recognized by the physical sciences. But this view raises new problems.[4] Does "recognized by the physical sciences" mean current science? What if science changes as it has done in the past? That would mean that today's physicalism would be false. Does it mean instead something like an ideal future science? If so, that is something we don't have access to. Physicalism at present would therefore not posit an ontology, which is prima facie bad for an ontological position. Although "naturalism" is not mentioned by Hossenfelder, it is on the same trajectory: the only things that exist are made up of the stuff studied by the natural sciences. Naturalism faces the same questions as physicalism.[5]

Reductionism is a different issue. (Several varieties of it will be discussed in Chapter 6.) Hossenfelder says "reductionism" means that the behavior of a macro-object can be "deduced from" facts about its constituent parts. (This is not in fact what philosophers generally mean by "reduced to," but let's ignore that for now.) Reductionism, thus defined, is a wildly implausible description of the whole of science. Does she think one can deduce how male penguins care for their young from truths about their constituent atoms? Since, of course, she can do no such thing, she qualifies the claim:

> We certainly know of many things that we cannot currently predict, for our mathematical skills and computational tools are limited. The average human brain, for example, contains about a thousand trillion trillion atoms. Even with today's most powerful supercomputers, no one can calculate just how all these atoms interact to create conscious thought. But we also have no reason to think it is not possible. For all we currently know, if we had a big enough computer, nothing would prevent us from simulating a brain atom by atom.
>
> (Hossenfelder 2022, 84)

So while the deduction from micro-constituent to macro-behavior cannot be done today, this is just a limitation on our mathematical knowledge and computational capacity—something we will overcome in time.

Hossenfelder goes on to deny free will in terms similar to Greene's: "[The] trouble with free will is that according to the laws of nature that we know describe humans on the fundamental level, the future is determined by the present" (2020). By "the present" she means that the physical entities within the universe each have a state at this instant in time. A state is roughly all the instantaneous values of a unique set of properties a system could have. For example, take the air molecules in a box. A state of this system could be specified by the value of momentum of each particle along with its position at a point in time. That's a lot of information. From a given state, the behavior of the molecules is dictated by the laws of mechanics. Physicists like Pierre-Simon Laplace believed we could extend the example to all the particles in the universe (Section 2.2.1.2). Hossenfelder believes the same idea applies to individuals:

> These deterministic laws of nature apply to you and your brain because you are made of particles, and what happens with you is a consequence of what happens with those particles... [You] can insist to just not believe reductionism is correct. But this is denying scientific evidence. We do not guess, we know that brains are made of particles. And we do not guess, we know, that we can derive from the laws for the constituents what the whole object does. If you make a claim to the contrary, you are contradicting well-established science. I can't prevent you from denying scientific evidence, but I can tell you that this way you will never understand how the universe really works.
>
> (2020)

Given that we are made up of particles, and the laws of nature determine the behavior of those particles, there is no place for free will. Anyone claiming otherwise is "contradicting well-established science." Philosophical distinctions are well and good, she says, but science has discovered the truth. Libertarians simply do not "understand how the universe really works." We might ask again, is it true that "we can derive from the laws for the constituents what the whole object does"? Can she derive the behavior of hemoglobin from quantum mechanics alone? Well no, but we can do so "in principle" (Hossenfelder 2019, 90)—a two-word phrase that does *a lot* of work here.

Hossenfelder is the clearest example of fundamentalism mentioned in the preface. I will say more about it in Chapter 6, but for now, I take fundamentalism to be the conjunction of three claims: (1) ontological priority of microphysical entities, (2) explanatory reduction to fundamental physics, and (3) epistemic privilege of fundamental theories. Sean Carroll offers a distinct argument for this third clause.

1.3.3 Sean Carroll

Carroll is the most philosophically informed of the physicists discussed in this chapter. As we have seen in Section 1.1, he argues that libertarian free will would require breaking the laws of nature. His argument is unique, so here is a rough analogy for what he intends to show. Say that someone claims there is an undiscovered planet in our solar system. We have detailed models of the orbits of the known planets, including how they gravitationally interact with one another. For there to be another planet, the orbits of the known eight would need to be different due to the gravitational pull of the undiscovered one. Given the orbits we observe and our understanding of gravity, there is no place in the solar system for an undiscovered planet. Carroll argues similarly that there is no place in physical reality for creatures with free will.

He begins with the idea that we now know all the forces and all the particles in nature almost down to the fundamental level (2016, 176). These are found in the Standard Model of particle physics. There might be more going on at still smaller scales, as string theory proposes, but there is little hope of exploring that level experimentally. The energies required to explore that realm are just too high—many orders of magnitude higher than CERN's Large Hadron Collider, which verified the existence of the Higgs boson. But even if we could do so, nothing at that deeper level would require changing the Standard Model. And it is the Standard Model, says Carroll, that provides everything we need for grounding the phenomena of the macro-realm:

> I will argue that we have good reason to believe that this model is both *accurate* and *complete* within the everyday-life regime; in other words, that the laws of physics underlying everyday life are, at one level of description, completely known. This is not to claim that physics is nearly finished and that we are close to obtaining a Theory of Everything, but just that one particular level in one limited regime is now understood. We will undoubtedly discover new particles and new forces, and perhaps even phenomena that are completely outside the domain of applicability of quantum field theory; but these will not require modifications of the Core Theory...
>
> (Carroll 2021, 2)

What Carroll calls the "Core Theory" is the Standard Model plus some special cases in general relativity.[6] Our degree of certainty in the physics of the Core Theory is very high. Whatever else we might discover at still smaller scales will not change it.

Here's an analogy for what Carroll has in mind. Consider the periodic table of elements. We know these elements are manifestations of the deeper physics and more exotic particles described by the Standard Model. But we also know that whatever new particles might one day be discovered, it's highly unlikely that this discovery would change anything in the periodic table. Freshman chemistry rests on a sure foundation. Nothing discovered at CERN will require any change in its textbooks. For example, nothing will be found that would require placing a new element between silver and cadmium. Silver has atomic number 47 and cadmium 48, which represent the number of protons in their respective nuclei. There can't be a 47½. Hence, there are limits on what might be discovered in the future.

Given what we currently know about particle physics, says Carroll, similar limitations exist on what might yet be known. Whatever yet-to-be-discovered entities reside down there, two things are possible: (i) the new particles/forces only exist at the much higher energies produced by supercolliders and then quickly decay away, or (ii) they are too weakly coupled to anything else to make a detectable difference. "Coupling" has to do with whether one thing can causally affect another. All bits of matter sense the pull of gravity, and so are coupled to the gravitational field. Neutrinos are electrically neutral and are not (directly) coupled to the electromagnetic field. According to the second possibility mentioned here, any unknown particles or forces will have little or no influence on the other entities in the Standard Model.

This puts limits on what a new discovery can mean for the macro-realm. "[Any] new particles or forces must interact too weakly with Core Theory fields to be relevant to everyday-life phenomena" (Carroll 2021, 3). So, while we might discover something new, whatever it is will not require any change to the Core Theory/Standard Model that would make a difference at the level of our experience. Again, he could have made the same point using chemistry. Whether strings exist or not, or whether new, short-lived particles are found in a supercollider, the periodic table will not change.

So what does this have to do with free will? Libertarian freedom, Carroll says, would require some new causal influence—something beyond the physics of the Core Theory, which he believes is deterministic.[7] Whatever that causal influence might be, it would somehow have to interact with the particles and forces in the Core Theory. That, he claims, is the problem. Given the experimental constraints on what is currently known, any undiscovered influence can be at most weakly coupled to anything in the Core Theory (Carroll 2021, 4). Even if there were some new substance or property out there that is currently hidden from science, it wouldn't be able to significantly affect the particles in the Standard Model. But if it cannot influence those particles, which we are composed of, then it cannot affect our actions. A new, causally significant source would be equivalent to element 47½ on the periodic table: it just can't be. Hence, Carroll believes that our best physics bars the door to the sources of action needed for libertarian freedom.

In short, everything in the macro-realm is made up of stuff in the Core Theory. The Core Theory has no space for anything new that could influence the entities of everyday experience, including our bodies. Since that sort of influence is necessary for libertarian free will, Carroll concludes that physics shows there is no such thing.

1.3.4 Fundamental Physics and the Macro-world

One might wonder why this line of thought only targets free will. Wouldn't the same idea apply to any causal claim in the macro-realm? For example, physicians say that statins lower cholesterol, but statins are not found in the Core Theory. If they are to *do* anything, says Carroll, they would have to somehow causally affect the particles in fundamental physics. But physics does not deal with pharmaceuticals or most other things that influence biological systems. If Carroll is correct, genes, cells, and organs would seem to be just as causally inert as libertarian freedom. Doesn't his argument apply to all causal claims outside of particle physics?

This, Carroll would rightly say, is a somewhat different issue. The question here is about the relation between fundamental physics and the other sciences. Hossenfelder and Carroll instead take chemistry, biology, and the rest to usefully describe the world in a coarse-grained sense. Thermodynamics has been an influential example. Experimentalists can easily measure the thermodynamic properties of a gas, such as temperature, pressure, and density. This allows for two levels of description for, say, the air in a bottle (Carroll 2016, 97). At the macro or "coarse-grained" level, there are the thermodynamic properties just mentioned. If more precise mathematical modeling is needed, most gases can be treated as continuous fluids, which is how they behave in aerodynamics. At the micro or "fine-grained" level, there is the massive number of particles described by statistical mechanics. While the macro is more convenient given our computational limitations, the two should track each other over time:

> One of the features of the molecules/fluid example is that we can derive the macroscopic fluid theory from the microscopic molecular theory. That is, we can start with the molecules, assume that there is a high density of molecules at every point in space, and then "smooth out" the distribution to obtain explicit formulas for fluid properties such as pressure and temperature in terms of what the molecules are doing.
>
> (Carroll 2016, 99)

In other words, there is a way to start with statistical mechanics and the microstate of the molecules in the bottle at some time t_1 and then derive its corresponding macro-state description at that time. The equations for the macro-state can then say how the system evolves from t_1 to some later time t_2. One can then derive the corresponding microstate at t_2. (There are, in fact, many microstates that correspond to any given macro-state.) From the physicist's point of view, there is one system with two different descriptions. Using the macro to see how the system evolves over time is a purely pragmatic decision. The macro "is enormously more computationally efficient than the microscopic molecular theory" (Carroll 2016, 99). If we had more precise data and computational capacity, thermodynamics would not be needed. All the calculations could be made in principle at the level of atoms. This is the basis for the claim that thermodynamics has been reduced to atomic physics.

As we will see in Chapter 6, this example has played an outsized role in thinking about the relation between macro- and micro-levels of description. Philosophers of physics, such as Mark Wilson, emphasize that the relation between thermodynamics and the atomic theory of gases is unique (2017, 230). Engineers and applied physicists working with ceramics, metals, and composites know that there is no handy macro-to-micro correspondence for their materials. Carroll himself goes on to say that the reduction of thermodynamics to statistical mechanics, which in many ways is just a kind of averaging over microstates, is a special case that "doesn't readily extend to other situations we might be interested in" (2016, 99).[8] Nonetheless, what happens at the macro level is determined by the micro level:

> If a macroscopic theory has a domain of applicability that is a subset of the domain of applicability of some microscopic theory, and both theories are consistent, then the microscopic theory can be said to *entail* the macroscopic one; but that's often something we take for granted, not something that can explicitly be demonstrated.
>
> (Carroll 2016, 100–101)

While he admits that our understanding of thermodynamics and statistical mechanics is somewhat unique, something similar must hold across the board even though the details are currently beyond us.

Why think that? There is no denying the power of statistical mechanics, but can we really generalize from this one example to the whole of science? While it can sometimes sound that way, there is more to the story. "[Physicists] now have a well-defined procedure to derive, say, the properties of water from the properties of atoms. This procedure is called *coarse-graining*, and while the mathematics is tough, the idea is conceptually simple" (Hossenfelder 2022, 83). What Hossenfelder has in mind is more than considering a single system from two different scales, like the molecules in the bottle. There the focus was on averaging the properties at a small scale to derive the properties at a larger one. Twentieth-century techniques went further:

> Coarse-graining in physics is a more complicated version of this averaging; it's a method for discarding information you don't need. In physics, the size of the grid is often referred to as the *cutoff*, and the task is to write down an approximate model that is accurate enough at the resolution given by the cutoff, plus small corrections for the missing details. If you then throw away the small corrections below the cutoff for good, you have what physicists call an **effective model**. This model

> is not fundamentally correct—because, like your averaged topographic map, it is missing information—but it is good enough at the level of resolution you are interested in.
>
> (Hossenfelder 2022, 84–85)

The justification is again pragmatic. The models produced are "good enough" for the purposes at hand. Even though the technique sketched here applies only to special cases,[9] Hossenfelder uses them to support her reductionism and the primacy of fundamental physics. As we will see in Chapter 8, some philosophers of physics believe the details are far more significant than she realizes.

Hossenfelder and Carroll take coarse-graining in some sense or other to be the basis for the relation between physics and the other, higher-level sciences. There is only one physical reality, and the non-fundamental sciences describe it in varying degrees of resolution, human psychology included. "There is no obstacle to that kind of vocabulary of desire and intentionality being 'true,' as long as its predictions are compatible with those of other successful vocabularies" (Carroll 2016, 373). Each level of the hierarchy of sciences can be useful and "true," so long as they are logically consistent with the truth at lower levels.

Carroll puts scare quotes around "true" when it comes to descriptions about the macro-realm. Why is that? Are well-established truths of psychology not really true? What about zoology or biochemistry? For many physicists, especially particle physicists, I believe the answer is that the rest of science is not true *simpliciter* since all the other levels employ terms and concepts not found at the level of fundamental physics. Macro-theories have a sort of pragmatic truth that sanctions their use but is still somehow lesser compared to the deep truths of the Standard Model. This means that physics trumps all other claims, including those involving free will.

While the scientists cited thus far do not mention it, there is one more physics-based objection to libertarianism. It proceeds in two stages. First, special relativity entails (or at least supports) a view in the philosophy of time known as *eternalism*. The eternalist believes that the past and the future are real. From a God's-eye point of view, the entire timeline exists. The dinosaurs are still there in some sense, as is the Sun in its red giant stage some five billion years after today. Moreover, that timeline is static and unchanging. The second stage is more controversial: this lack of change means that we do not have free will. What we think of as the future is fixed, and nothing one does now can change it. Many believe that the ability to change the course of the future is a necessary condition for free will. Relativity therefore poses a challenge to libertarianism.

The main reason this book focuses on critiques from physicists is that their views have an outsized role in shaping both popular and scholarly opinions. Roughly the same arguments presented in this section can be found in philosophy as well.

1.4 Philosophers Piling On

There are many examples of scientists embracing reductionism, which is understandable. Part/whole thinking has had tremendous utility in the history of science. The existence of atoms was widely accepted well before there was direct empirical evidence. It's not surprising, then, that particle physicists lean into reductionism.

Philosophers, on the other hand, have broader interests, so one might expect them to express a more nuanced view. This is typically not the case. Consider Derk Pereboom:

> Can agent-causal libertarianism be reconciled with what we would expect given our best physical theories? If this theory is true, then when we make free decisions, we cause these decisions without being causally determined to do so. As agent-causes that fit this description, we at some point would affect the physical world external the agent-cause, perhaps most directly in the brain. If the physical world were generally governed by deterministic laws, it seems that at the point of interaction we would then encounter divergences from these laws.
>
> (Pereboom 2007, 111)

We can ignore what "agent-causal" libertarianism amounts to for now.[10] Pereboom's point is the same as Carroll's: "In order for libertarian freedom to exist, it would have to be possible for human beings to overcome the laws of physics just by thinking." In the exercise of libertarian freedom, what would have happened according to the laws does not happen, which is impossible.

Even if nature were ultimately indeterministic, there would still be statistical laws at work. For example, radioactive decay is an indeterministic process, and yet we know the half-life of uranium-235 is 703.8 million years. This allows one to calculate the probability of a given uranium-235 atom decaying in the next day, century, etc. If a given lump of uranium did not undergo decay, it would run counter to Schrödinger's equation and Born's rule, which provide the relevant probabilities. It would thus violate the laws of nature. Similarly, if there were creatures with libertarian freedom, free choices would diverge at least some of the time from whatever probabilistic laws are in place (Pereboom 2007, 112–13). As philosopher Barry Loewer concludes,

> Neither deterministic versions of quantum mechanics like Bohm's theory nor objective indeterministic theories like GRWP make room for

> libertarian freedom. It appears that if libertarian freedom is coherent at all, it must be that free choices ... are outside of the nomological order.
> (Loewer 1996, 106)

That is a high price. Being "outside of the nomological order" is roughly equivalent to being supernatural. While I am not a naturalist and so find the possibility of events that are not law-governed acceptable, if one has to believe in the supernatural to accept libertarianism, it will be far less attractive.

Reductionism of one sort or another is common among philosophers. They are also more likely than physicists to raise a closely related issue: the causal closure of physics (CCP). Closure says that if an event at the level of physics has a cause, that cause is also physical. Whatever might be going on at the level of consciousness and will, those events cannot make a causal difference at the level of fundamental physics unless they *just are* manifestations of events at that level. Carroll's remarks about possible new forces and the Core Theory dovetail nicely with CCP. Even if free-will decisions exist, they cannot influence the particles that compose the agent's body.

1.5 What Might a Libertarian Say?

These are serious objections to libertarian freedom. The fact that they are grounded in physics sets them apart from the normal back and forth of many philosophical debates. When well-known physicists agree that science forbids free will, it carries significant weight. What might a libertarian say in reply?

As a preview of what's to come, note again that Greene and Hossenfelder hold a kind of part-whole reductionism. Their argument runs roughly as follows:

> Human bodies have constituent parts, namely atoms.
> The behavior of atoms is wholly determined by the laws of physics.
> Hence human bodies are wholly determined by the laws of physics.

Is this a good argument? Let's compare it to another:

> Human bodies have constituent parts, namely atoms.
> Atoms are colorless.
> Hence human bodies are colorless.

Clearly that doesn't work, but what has gone wrong? Several things. First, both arguments commit the fallacy of composition, but that's not all that interesting.[11] Recall this passage quoted earlier:

> We certainly know of many things that we cannot currently predict, for our mathematical skills and computational tools are limited. The average human brain, for example, contains about a thousand trillion trillion atoms... But we also have no reason to think it is not possible. For all we currently know, if we had a big enough computer, nothing would prevent us from simulating a brain atom by atom.
>
> (Hossenfelder 2022, 84)

Hossenfelder believes that human psychology can be reduced to brain activity, and brain activity can be reduced to the behavior of its constituent atoms. While we currently lack the technology to simulate a brain at the atomic scale, there is no reason to think it's impossible.

Most philosophers a century ago would have agreed. Although science has gaps and discontinuities, especially across disciplines, they believed that eventually these will all be resolved. In time it will be clear how psychological states are produced by brain states, brain states by biochemistry, and on down the line without holes or handwaving.

Many philosophers of science now see this picture of fully unified scientific knowledge as overly optimistic. It is not that we merely lack some crucial data or better computational techniques. Instead, the idea that one day the truths in all social, biological, and physical sciences will fit together in a neat hierarchy has largely been abandoned. At present, not even the various subdisciplines within physics match this description. Hossenfelder's hope that exhaustive knowledge of atoms plus smarter mathematicians plus better computers will overcome these limitations is at best naïve. She jumps from a mereological truth—that we and all other macroscopic entities are composed of atoms—to the far stronger claim that scientists will eventually be able to predict the behavior of macroscopic objects from knowledge of those constituent parts. While atomism *is* "one of the best-established facts about nature," the behavior of atoms entails very little about the behavior of the macroscopic realm. As we will see, most of what is going on in the micro-realm is irrelevant to the world of our experience. Hossenfelder acknowledges this,[12] but does not recognize its significance. If she did, it would undermine her faith in reductionism.

Fortunately, Hossenfelder tells us what it would take to show her wrong. "[If] you wanted to prove reductionism false, you'd have to show that describing a system in macroscopic terms results in different predictions than those you'd get from its microscopic description ..." (2022, 89). She believes it is a safe bet that this will never happen. As we will see in Chapter 8, however, there are many examples where relying solely on microphysics would lead to false predictions of the macro-realm. This is because the micro-parts alone do not fully determine the behavior of the macro-whole in many systems.

In the end, I will show that libertarian free will is not incompatible with physics. The conflict is not with physics itself but rather physics wedded to two or three questionable philosophical assumptions. Those assumptions are what do the heavy lifting on this issue, not the physics. The more ambitious goal is to say what sort of ontology of agents is required for libertarianism. What sorts of beings might exercise such freedom? There are two answers. The first is mind-body dualism. This is the easier case, at least at first glance. Immaterial minds are not governed by the laws of nature and so are not subject to causal determinism grounded in such laws. This is one reason libertarians are often more favorably disposed toward dualism. "Wait," says the physicalist, "that doesn't help. Dualism itself violates causal closure and the laws of nature, principally the conservation of energy." Closure, I will argue, is one of the questionable philosophical assumptions mentioned here. As for violating conservation laws, that would indeed be a "fatal flaw" (Dennett 1991, 35) if only it were true. Dualism does not violate conservation of energy, and understanding why it does not is interesting and instructive.

Dualism is, however, not a popular option in today's intellectual landscape. If libertarianism requires one to embrace dualism, that would be too high of a cost for many. Fortunately, philosophy of science has become less reductionist, at least in some corners. It now offers the resources to make a case for emergence, including emergent free will. Before offering that argument, we will have to map the various types of reduction and emergence in what has become a very messy terrain. (If "emergence" is unfamiliar, take it for now as the opposite of reduction. Emergent properties cannot be reduced to their constituent entities.)

The ultimate goal was announced in the preface: to refute the claim that "libertarianism ... [is] incompatible with what we know about nature" (Hossenfelder 2022, 128). Is this book only for libertarians, then? No, as we'll see in the next chapter, there are other types of free will. But if we can show that physics does not forbid libertarian freedom, then all others are safe as well.

1.6 Overview

Here is an overview of the rest of the book.

Chapter 2: What Is Free Will? I have been using terms like "libertarian" somewhat loosely thus far. In this chapter, we start with the distinction between determinism, libertarianism, and compatibilism. There are several reasons to believe that nature is deterministic and different types of determinism that could be a threat to free will. Libertarians typically agree that if the case for determinism could be proved, then there is no free will. Compatibilists believe otherwise. As the name implies, compatibilists hold that free will and determinism need not be in conflict with one another. We will also consider more recent distinctions between source and leeway notions of freedom as well as event and agent causation. Chapter 2 makes clear what it is the free-will skeptics are taking aim at (libertarian freedom) and the main threat any reply must overcome (determinism).

Chapter 3: The Challenge of Determinism. While determinism is the main obstacle that libertarians must overcome, physical determinism is a more complex topic than it seems on the surface. We need to understand precisely what it entails and its historical development to see how the tension with free will arose. When it comes to physics, the conventional wisdom is that classical mechanics was deterministic, but quantum mechanics is not. Both claims may in fact be false. Once the details are in hand, a relatively easy way to escape physical determinism presents itself: mind-body dualism. Immaterial minds are not subject to the laws of nature. A second, more difficult approach involves emergence, which will take up the last three chapters of the book.

Chapter 4: Relativity and the Block Universe. This chapter is a bit of an interlude, discussing a different sort of challenge to free will posed by special and general relativity. Many believe that relativity entails a view of time in which the future fully exists. It is, just like the past, fixed and unchanging. Moreover, this lack of change means that we do not have it in our power today to make a choice that would alter the course of future events, which, as I note in Chapter 2, many libertarians take to be a necessary condition for free will. If there is an unchanging truth that you will have pizza for dinner next Friday, in what sense are you free not to do so? We will investigate whether relativity genuinely entails a fixed future or whether alternative interpretations of the physics can accommodate the kind of open future that libertarian freedom requires.

Chapter 5: Physics vs. Dualism. This chapter returns to the idea that mind-body dualism presents a way of escaping physical determinism. Dualism itself is, of course, controversial. We will examine two physics-based objections here that have already been mentioned. The first is CCP. If closure is true, then even if minds exist, they cannot influence those entities recognized by physics. The second is conservation of energy. For a mind to influence a brain, it would have to be a source of energy, which conservation forbids—or at least it does according to a freshman-level understanding of conservation laws. In the end, dualism survives both objections. The unresolved problems are philosophical. Is one willing to take on the ontological commitment of an immaterial agent? Most philosophers will not, and so I will push on to another approach that will take substantially more work to make plausible. Before that, however, we will take a brief detour.

Chapter 6: Reductions. Many of the arguments in Chapter 1 are rooted in a reductionist approach to science. What does that mean? Several different things, it turns out, that are often jumbled together. Sometimes it is simply that all material stuff is composed of atoms. Other times it has to do with how terms in one scientific discipline relate to those in another. We will consider a brief taxonomy of views and see why the scientific case for reductionism is less compelling than it might appear. This critique of reductionism establishes why an alternative approach like emergence is needed and justified.

Chapter 7: Emergence. Physicists and philosophers have more recently warmed to the notion of emergence, but the term is used in many ways. This chapter distinguishes between weak emergence, which emphasizes epistemological limitations in predicting higher-level phenomena from lower-level properties, and strong emergence, which posits novel causal powers at higher levels that cannot be reduced to their base. Two more recent approaches are contextual emergence, which highlights the role of environmental conditions in shaping system behavior, and weak metaphysical emergence, which argues for the autonomy of higher-level entities by eliminating degrees of freedom. The chapter provides a context for the remainder of the book.

Chapter 8: Micro, Macro, and Meso. A positive case for emergence that provides space for genuine human agency begins here. Moderate metaphysical emergence (MME) is a concept grounded in physics, particularly continuum theories and condensed matter physics, that reconciles the autonomy of higher-level phenomena with the constraints of fundamental physics. Unlike weak emergence, MME takes emergent properties to be metaphysically real and causally relevant while avoiding the brute ontological commitments of strong emergence. The chapter demonstrates how systems with different microstructures exhibit universal macro-behaviors, such as phase transitions, which cannot be reduced to their micro-level constituents. A mathematical technique known as *renormalization* reveals that causally relevant properties often reside at meso-scales rather than the micro level, a principle further supported by multiscale modeling in engineering and biology. The chapter also rebuts the objection that emergence is merely pragmatically useful but does not tell us anything about nature.

Chapter 9: Compatibilist Libertarianism. Compatibilism and libertarianism are presented in the free-will literature as mutually exclusive positions. Christian List argues that one can hold both (2019). He's right. Unfortunately, his arguments have not gained much traction. By incorporating a historically informed understanding of determinism (Chapter 3), a dynamic view of spacetime (Chapter 4), a new model of emergence (Chapter 8), and an account of downward causation (Chapter 9), List's "compatibilist libertarianism" can be put on a firmer foundation.

List himself acknowledges that his account would need to be filled out by "a community of scholars, across multiple disciplines" (2019, 10). This book is part of that effort. Insights from MME show how the necessary conditions for free will (intentional action, alternative possibilities, and mental causation) can be met even if determinism is true at the lowest levels of reality. This provides a way to answer the free-will skeptics that is—like their critique—grounded in physics.

Chapter 10: The Big Picture. This chapter traces the arc of the book, starting with the skeptical challenges in Chapter 1. The only unresolvable conflict is not between libertarianism and physics, but rather libertarianism and a fundamentalist approach to science.

At the end of the day, libertarians should have no fear of determinism at the level of fundamental physics, if indeed those theories turn out to be deterministic. There is far more to reality than the Standard Model. Science provides ample space for libertarian freedom once physicalist assumptions have been disentangled from physics itself.

Notes

1 The Abrahamic faiths—Judaism, Christianity, and Islam—trace their origin to the patriarch Abraham.

2 See, for example, Hobbes in *Leviathan*, chap. 6.

3 Roughly from the mid-seventeenth to the early twentieth centuries.

4 What follows is known as "Hempel's Dilemma" (Hempel 1969).

5 And has no good answers, according to Michael Rea (2002). It should therefore not be construed as an ontology.

6 This is a partial reconciliation of quantum mechanics and general relativity. There is currently no theory that fully reconciles the two.

7 Carroll rejects interpretations of quantum mechanics that involve a collapse of the wave function. If he is correct, then determinism would be restored in the quantum realm, something discussed more fully in Chapter 3. This is not as important as some might believe. Ontological randomness in quantum mechanics has little bearing on the question of free will.

8 This is well known. "Direct derivational connections between, say, continuum theories of many-body systems and fundamental theories of their atomic constituents, are rare, and depend upon the special nature of the systems. Specifically, direct connections depend upon the systems being quite homogeneous in their lower-scale makeup" (Batterman 2021, 138).

9 "In most cases, we currently cannot perform the mathematical calculations that would be necessary for coarse-graining" (Hossenfelder 2022, 86).

10 We will pick it up again in Section 2.4.3.

11 To be picky, the two arguments do not have the same logical form since the first is about the behavior of atoms and the second is about one of their properties. If that is bothersome, change the second premise to "Atoms have the property of being wholly determined by the laws of physics" and the conclusion to "Hence

human bodies have the property of being wholly determined by the laws of physics."

12 "Why is it that the details from short distances do not matter over long distances? Why doesn't the behavior of protons and neutrons inside atoms matter for the orbits of planets? How come what quarks and gluons do inside protons doesn't affect the efficiency of drugs? Physicists have a name for this disconnect—the *decoupling of scales*—but no explanation. Maybe there isn't one" (Hossenfelder 2022, 89).

Works Cited

Augustine. 1887. *City of God*. Edited by Philip Schaff. Translated by Marcus Dods. Nicene and Post-Nicene Fathers. Vol. 2. Buffalo: Christian Literature Publishing. www.newadvent.org/fathers/120105.htm.

Batterman, Robert W. 2021. *The Middle Way: A Non-Fundamental Approach to Many-Body Physics*. Oxford: Oxford University Press.

Carroll, Sean. 2016. *The Big Picture: On the Origins of Life, Meaning, and the Universe Itself*. New York: Dutton.

Carroll, Sean. 2021. "The Quantum Field Theory on Which the Everyday World Supervenes." *arXiv:2101.07884* [physics.hist-ph]. https://arxiv.org/abs/2101.07884.

Dennett, Daniel. 1991. *Consciousness Explained*. Boston: Little, Brown and Co.

Fergusson, David. 2019. *The Providence of God: A Polyphonic Approach*. Current Issues in Theology. Vol. 11. Cambridge: Cambridge University Press.

Greene, Brian. 2020. *Until the End of Time: Mind, Matter, and Our Search for Meaning in an Evolving Universe*. New York: Alfred A. Knopf.

Hasker, William. 2004. *Providence, Evil, and the Openness of God*. New York: Routledge.

Hempel, Carl. 1969. "Reduction: Ontological and Linguistic Facets." In *Essays in Honor of Ernest Nagel*, edited by Sidney Morgenbesser, Patrick Suppes, and Morton G. White, 179–99. New York: St. Martin's Press.

Hossenfelder, Sabine. 2019. "The Case for Strong Emergence." In *What Is Fundamental?*, edited by Anthony Aguirre, Brendan Foster, and Zeeya Merali, 85–94. The Frontiers Collection. Cham: Springer International Publishing. https://doi.org/10.1007/978-3-030-11301-8_9.

Hossenfelder, Sabine. 2020. "You Don't Have Free Will, but Don't Worry." *BackReAction* (blog). October 10, 2020. http://backreaction.blogspot.com/2020/10/you-dont-have-free-will-but-dont-worry.html.

Hossenfelder, Sabine. 2022. *Existential Physics: A Scientist's Guide to Life's Biggest Questions*. New York: Viking.

Laplace, Pierre Simon. (1814) 1902. *A Philosophical Essay on Probabilities*. Translated by F. W. Truscott and F. L. Emory. Chichester: Wiley.

List, Christian. 2019. *Why Free Will Is Real*. Cambridge: Harvard University Press.

Loewer, Barry. 1996. "Freedom from Physics: Quantum Mechanics and Free Will." *Philosophical Topics* 24 (2): 91–112.

O'Keefe, Tim. 2021. "Ancient Theories of Freedom and Determinism." In *Stanford Encyclopedia of Philosophy*, edited by Edward N. Zalta. Stanford: Metaphysics Research Lab, Stanford University. https://plato.stanford.edu/archives/spr2021/entries/freedom-ancient/.

Pereboom, Derk. 2007. "Hard Incompatibilism." In *Four Views on Free Will*, edited by John Martin Fischer, 85–125. Great Debates in Philosophy. Oxford: Blackwell.

Rea, Michael C. 2002. *World without Design: The Ontological Consequences of Naturalism*. Oxford: Oxford University Press.

Timpe, Kevin, and Jonathan D. Jacobs. 2015. "Free Will and Naturalism: How to Be a Libertarian and a Naturalist Too." In *The Blackwell Companion to Naturalism*, edited by Kelly James Clark, 319–35. Chichester: Wiley-Blackwell.

Truesdell, Clifford. 1984. *An Idiot's Fugitive Essays on Science: Methods, Criticism, Training, Circumstances*. New York: Springer-Verlag.

Wilson, Mark. 2000. "The Unreasonable Uncooperativeness of Mathematics in the Natural Sciences." *The Monist* 83 (2): 296–314.

Wilson, Mark. 2017. *Physics Avoidance: Essays in Conceptual Strategy*. Oxford: Oxford University Press.

2 What Is Free Will?

2.1 The Lay of the Land

If you are not interested in free will, you should probably stay away from philosophy. Aristotle discussed it, as has virtually every major philosopher since. The website PhilPapers.org lists over 4,500 articles and book chapters currently in print dealing with every conceivable aspect of the question. Any overview—which this chapter is intended to be—will inevitably ignore large chunks of the literature, and so I make no apologies for doing so. One must find ways to narrow the focus. How will that be done here?

While there are many challenges to the notion that we have free will, I am most interested in those posed by physics. What counts as a central question as opposed to a rabbit trail will be determined by whether the problem is grounded in physical science. Of course, one person's rabbit trail is someone else's dissertation topic. When forced to choose, we will peer down the path just a bit, then quickly get back to the main topic.

We start with the basic distinction between libertarian freedom, determinism, and compatibilism. The first is so intuitively obvious to many that it seldom needs a defense, at least initially. Determinism, on the other hand, requires more work, and we briefly consider three reasons for thinking that reality is deterministic. Both libertarians and determinists typically believe that the two are incompatible with one another. If one is true, the other must be false. Compatibilism attempts to show that the two can be harmonized after all. We then move on to two distinctions: *leeway* and *sourcehood* approaches to libertarianism, and *event* as opposed to *agent* accounts of causation.

Some of the material in this chapter will be familiar to readers with even passing knowledge of the free-will literature. It is intended primarily for students and those coming to the topic with a background other than philosophy.

DOI: 10.4324/9781003681199-2

2.2 Three Main Views

Let's begin with the standard three-part distinction.

2.2.1 Determinism

The first view is straightforward. Determinism says that regardless of our sense that we can change how events will unfold, there is in fact only one set of outcomes possible, only one way that the universe can evolve. That feeling of control that you and I share is merely an illusion, says the determinist. The truth is that there is simply no space for the exercise of free will and no place among the causal joints of the cosmos in which one's free choices can make a difference. Determinists know that may sound depressing, but if you wish to deal with reality rather than psychological projections, they think you should just get over it. In the end, it might not be as bad as you think (Pereboom 2007, secs. 10–12).[1]

There are several roads that lead to determinism. Let's start with one that predates modern science.

2.2.1.1 Theological Determinism

One of the theistic God's essential properties is omniscience, and most theists believe this includes perfect knowledge of the future.[2] Suppose that today God believes that Phil will have pizza for dinner next Saturday. Does Phil have the ability to eat something other than pizza next Saturday? Say that Phil has a cheeseburger instead. That would mean that God has a false belief today, which is incompatible with omniscience. If God is omniscient, then it seems that Phil cannot have a cheeseburger next Saturday after all. God's perfect foreknowledge is in tension with our ability to do other than what God currently believes will come to pass. There is only one way that the future will unfold; God knows precisely what it is, and we cannot change it. An entirely different book could be written about how philosophers and theologians have tried to resolve this tension (Hasker 2011; Zagzebski 2021).[3]

Historically, theological determinism was more closely associated with a strong doctrine of providence than foreknowledge, a link most clearly forged by Augustine (Fergusson 2019, 48–51) and fully embraced by Reformed Protestants. Schaff ([1563] 1877) defines providence this way:

> The almighty and ever present power of God by which God upholds, as with his hand, heaven and earth and all creatures, and so rules them that leaf and blade, rain and drought, fruitful and lean years, food and

> drink, health and sickness, prosperity and poverty—all things, in fact, come to us not by chance but by his fatherly hand.
>
> (Question 27)

God's sovereign, providential care extends down to every event. All that happens is ultimately under God's direction, including the conditions that lead to each person's decisions. By controlling such conditions, God is able to bring about the outcome of elections, wars, etc. Medieval and early modern theologians embraced this doctrine as a source of comfort. No matter how bad things are, a loving God is ultimately in control. Such meticulous control is at odds, however, with libertarian freedom. Many theists then and now therefore reject libertarianism (Pereboom 2016), although a different sort of free will might be preserved (Section 2.2.3).

2.2.1.2 Physical Determinism

The skeptics in Chapter 1 had a different sort of determinism in mind, one that involves causation and the laws of nature. Under physical determinism, also known as "causal determinism," every event has a cause or set of causes that bring that event about.[4] Nothing in the world "just happens." Such a view reached its peak in the age of classical mechanics. Consider a simple system like a game of tug-of-war. The movement of the flag is obviously caused by the relative strength of the two forces acting along the rope. If we replace the human teams with pulleys and weights, the outcome can be perfectly calculated. However the flag moves is how it must move given the forces involved and the laws of mechanics. The same principle applies to more complex systems, from atoms and billiard balls to ships and solar systems. Once calculus was coupled to the idea that nature has laws, it became common to talk about laws of motion and the instantaneous *state* of a system, the overall configuration of the system at a point in time. The initial state of a billiard table, for example, would include the mass and positions of each ball, all at rest. While different systems require different variables to describe a state, if a deterministic system starts from some state, it will evolve in a particular way. If the system is reset to precisely the same state, it will evolve in the same way. In classical mechanics, the system must behave that way given the laws of nature.[5] That's just what it is for the system to be deterministic.

An illustration of this was presented by Pierre-Simon Laplace in the nineteenth century. While deterministic laws allow for successful predictions in science, we are limited in both knowledge and computational capacity. The behavior of complex systems is hard to predict. But say that a superintelligence did know all the relevant laws and the state of every particle in the universe at a given instant in time. Such a being could correctly calculate the state of the universe at any point in time. Laplace believes this follows from physical determinism:

> We ought then to regard the present state of the universe as the effect of its anterior state and the cause of the one which is to follow. Given for one instant an intelligence which could comprehend all the forces by which nature is animated and the respective situation of the beings who compose it—an intelligence sufficiently vast to submit this data to analysis—it would embrace in the same formula the movements of the greatest bodies of the universe and those of the lightest atom; for it, nothing would be uncertain and the future, as the past, would be present in its eyes.
>
> ([1814] 1902, 4)

Such an intelligence is now called a "Laplacian demon." It is an idealization that not even modern supercomputers approximate. The entire thought experiment is often portrayed as a direct implication of classical mechanics, but Laplace himself never makes this connection. He did not believe in determinism because of physics. His determinism was based on metaphysical principles.[6] Nonetheless, the predictions of a Laplacian demon would be possible in a Newtonian world because its evolution is governed by deterministic laws. That, at least, is the conventional wisdom on the matter. We will consider how accurate that portrayal is in the next chapter.

But wait, why are we talking about classical mechanics? "We don't believe in classical mechanics anymore!" says the annoyed reader, "It's been refuted by quantum mechanics, which is indeterministic!" Well, maybe. Quantum mechanics has superseded classical mechanics, and under the standard "Copenhagen" interpretation, it is indeterministic. Two identical quantum systems placed in the same state can evolve differently over time. However, the standard view has come under pressure of late. There are two well-known interpretations that deny quantum indeterminism. One is Bohmian mechanics, named after physicist David Bohm (Bohm 1952; Goldstein 2017, sec. 3). The other is the many-worlds interpretation, first considered by Hugh Everett and then developed by Bryce DeWitt (Saunders 2010; Wallace 2012). While we do not need to consider the details here, the three interpretations—Copenhagen, Bohm, and Everett—are empirically equivalent. That means that the data are consistent with all three.[7] If either Bohmian mechanics or the many-worlds interpretation is correct and physical determinism is restored, then a new and improved version of the Laplacian demon could predict the future state of the universe.

There is another reason why quantum mechanics is not necessarily decisive when it comes to determinism. Even if indeterminism can be found at the level of fundamental physics, it is difficult to find at the level of everyday objects. This is why engineers tend to learn so little quantum mechanics. Quantum indeterminism cannot help the libertarian get around physical determinism unless indeterministic quantum events can somehow be amplified into the macro-realm—the level of our experience (Pereboom 2022, 8–9). While such amplification is theoretically possible (Hodgson 2011, 73–78), it is not where most libertarians are placing their bets.

2.2.1.3 Spacetime Determinism

As we will see in Chapter 4, another version of determinism is supported by special and general relativity. Prima facie, the two imply a view of time known as *eternalism*, sometimes called the "spatialization" of time. Just as the whole of space currently exists even though we only have epistemic access to a small part of it, eternalism says that the whole of time exists. According to eternalism, the past, present, and future are all equally real. This is rather different from the commonsense view that takes past/present/future to represent a metaphysically important distinction, i.e., there *is* something special about the moving point in time we think of as "now." Eternalism instead denies there is any one place on the timeline, such as "the present." All of time fully exists and cannot be changed.

This last claim is where determinism comes in, although a different sort of determinism than in the last section. A fixed future would seem to conflict with libertarianism. Consider Phil again, who is planning on having pizza for dinner next Saturday. From an eternalist point of view, what we are calling "next Saturday" is just as real as what we think of as "the present." That means that there is a truth, a fact-of-the-matter, about what Phil is eating on the date corresponding to next Saturday. More importantly, that fact cannot be changed. Phil may feel today that he might or might not have pizza next Saturday, but every event next Saturday is part of the fabric of reality, says the eternalist, and so cannot be removed or altered. Our inability to change the course of the future looks like determinism, although, as we will see in Chapter 4, some eternalists reject this claim.

2.2.1.4 Neuroscientific Determinism

Much of the recent discussion about free will focuses on neuroscience, psychology, and other social sciences (Nahmias 2010; Sapolsky 2023; Wegner 2017). The work of neuroscientist Benjamin Libet often takes center stage (Shepherd 2017). His experiments seemed to show that the brain activity responsible for a given action begins *before* a subject consciously decides to act.

> [Subjects] whose brains were monitored by electroencephalogram (EEG) were asked to flex their right wrists whenever they wished and not to plan their wrist flexes. When they did not report planning, then the ramping up of EEG activity, the "readiness potential," *preceded* the average reported time of conscious awareness of the decision … Libet claimed that decisions when to flex were made unconsciously at the earlier of these two times (Libet 1985, 536). Siong Soon [et al.] … summarized these findings and claims as follows: "Because brain activity in the [supplementary motor area] consistently preceded the conscious decision, it has been argued that the brain had already unconsciously made a decision to move even before the subject became aware of it."
>
> (Pereboom 2022, 39–40, italics added)

The conscious decision to act comes after one's brain begins the causal sequence for activating the nerves controlling the relevant muscle response. The earliest interpretation of these results was that conscious decision-making is merely an effect of brain activity, and a lagging one at that. The neural processes responsible for choice are temporally and causally prior.

There are two reasons I will not be pursuing those arguments further. First, as I said earlier, the free-will literature is too broad to completely address in one book, and I have chosen to focus on challenges from physics rather than the whole of science. Second, I do not find these objections to free will to be particularly worrisome. Philosophers Alfred Mele (2009) and Timothy O'Connor (2009b) have effectively rebutted the arguments based on Libet's experiments. They show that the results either do not have the ramifications for free will that many first believed or, at the very least, such interpretations are underdetermined by the data. This reassessment has even filtered down to periodicals like *The Atlantic* (Gholipour 2019). The recent, widely reviewed book by Stanford neuroscientist Robert Sapolsky, *Determined: A Science of Life without Free Will* (2023), comes across as surprisingly naïve to those familiar with the literature discussed in this chapter (Fischer 2023).[8]

Let's now move to the opposite end of the spectrum when it comes to free will.

2.2.2 Libertarianism

I tell students that libertarian freedom is what you believe you have until some philosopher or scientist talks you out of it. That's a joke, but it is not too wide of the mark. Nearly everyone feels that the future is open, that they have live choices before them, and that they typically have the ability to choose one option over the other. The choice is not something forced upon them, nor has it been determined in some way before the time the decision is made. For most situations in life, whatever one chooses to do, the libertarian believes one could have done otherwise (Clarke 2011, 330).[9] Here "doing otherwise" is a capacity that an agent has to set in motion one chain of events instead of another. At this moment, I can either get some more coffee or keep typing. This is true even if my mug is empty and I am highly predisposed to get more. As a working definition, let's say that libertarian free will is the capacity of an agent to make choices and perform actions that are, to a significant degree, up to them, in the sense that their choices and actions are not fully determined by factors beyond their control (such as prior events and the laws of nature).

Until recently, libertarians would often picture the future as branching into the future (Figure 2.1). In my example, one branch represents my getting more coffee, the other my dutifully continuing to work. Which way the future unfolds is at least partly up to me. Why think so? Primarily because it is so intuitively obvious. While I chose to keep writing, I could have done otherwise.

Not all libertarians today believe that the ability to do otherwise is essential to free will. A minority view is that moral responsibility is the central concern and free will is whatever it is that allows agents to influence events

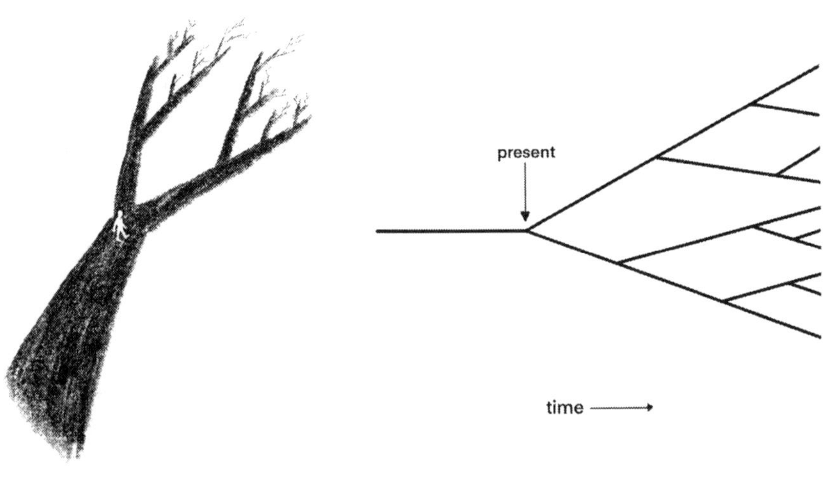

Figure 2.1 Forking paths.

Source: Artwork by Ashley McNutt-Greene. Used with permission.

such that they are morally responsible for those events (Timpe 2012, 5–8) (Pereboom 2022, 2). That moral responsibility requires free will seems clear. One is not responsible for events that have no causal connection to one's choices.[10] How an agent might have moral responsibility without being able to do otherwise will become clearer in Section 2.3.2.

One thing that determinists and libertarians have traditionally agreed upon is that the two are mutually incompatible. The third of the three standard views attempts to chart a path between these extremes.

2.2.3 Compatibilism

As the name implies, compatibilists hold that free will and determinism are compatible after all. How so? One way is by rejecting the idea that freedom requires the future to have an open, forking-path structure. So, yes, free will and determinism conflict if we accept traditional libertarian[11] freedom, but there are other ways to think about it. For early compatibilists, the key to freedom was a lack of constraint (McKenna and Coates 2021, sec. 2.1). If Al ties Phil to a chair, then Phil is no longer free to stand up. But when Phil is not so constrained, when he is acting according to his own internal desires, then he should be regarded as free. This is so even if, as many compatibilists believe, those desires are fully determined by physics, psychology, or other factors. In other words, they believe that determinism is (or may be) true at some level, but we still have free will if freedom is defined as acting on one's own desires rather than being subject to external coercion or constraint.

But this will not do, says the libertarian. No view that grants physical determinism can rightly be called free will, as Peter van Inwagen's much-discussed Consequence Argument tries to show (van Inwagen 1989; Campbell 2017). It starts by granting that every event in the past is a consequence of the laws of nature combined with the initial state of the universe (or the state just after the Big Bang), which is one way of understanding physical determinism. Of course, we have no influence over the primordial state of the universe or the laws of nature. The past is fixed and cannot be changed. Moreover, events in the present are consequences of these same laws plus the initial state of the universe. We therefore have no ability to influence those events in any way, including events involving our mental states and actions. But this conclusion applies to the future just as much as the present. The laws plus the initial conditions determine those events as well. The bottom line is that we do not have the power to influence how future events will evolve. Whatever trajectory the laws of nature determine for how events will unfold, we have no ability to change those events. But if we lack such a capacity, then there is no free will regardless of what compatibilists claim.[12]

Compatibilists have had much to say in response (Timpe 2012, 26–31; Capes 2019; McKenna and Coates 2021, sec. 4). Some argue that the future is not fixed but rather conditional. John Martin Fischer explains:

> This compatibilist analysis of freedom ... is called the "conditional analysis" because it suggests that our freedom can be understood in terms of certain conditional statements ("if – then" statements). More specifically, the conditional analysis commends to us the view that an agent *S*'s freedom to do *X* can be understood in terms of the truth of a statement such as, "If *S* were to choose (will, decide, and so forth) to do *X*, *S* would do *X*."
>
> (2007, 49)

For example, the antecedent might have to do with desire. If *S*'s strongest desire at a point in time were to eat a cheeseburger and a cheeseburger were available, *S* would eat a cheeseburger. If *S* had other desires, *S* would act differently. Instead of the strongest desire, the conditionals might involve, say, the best reason for acting at a point in time (Pereboom 2022, 19). In either case, the future depends on the antecedents, and so there is a sense in which, whatever *S* does, things might have gone differently.

Libertarians instead reject compatibilist freedom as lacking some necessary component. If the only way to have "chosen otherwise" is if the conditions were different—if *S* hadn't felt as hungry as he did, then *S* wouldn't have eaten that cheeseburger—that's not free will. Consider "If the football had the right momentum at a point in time, the football would go through the goalposts." While true, no one thinks this conditional has anything to do with free will. The agent has to have some ability to bring about (or prevent) the antecedent.

The most serious challenge to older versions of compatibilism is the so-called Manipulation Argument. Here is Richard Taylor's example:

> [Suppose] that an ingenious physiologist can induce in me any volition he pleases, simply by pushing various buttons on an instrument to which ... I am attached by numerous wires. All the volitions I have in that situation are, accordingly, precisely the ones he gives me. By pushing one button, he evokes in me the volition to raise my hand; and my hand, being unimpeded, rises in response to that volition. By pushing another, he induces the volition in me to kick, and my foot, being unimpeded, kicks in response to that volition. We can even suppose that the physiologist puts a rifle in my hands, aims it at some passerby, and then, by pushing the proper button, evokes in me the volition to squeeze my finger against the trigger, whereupon the passerby falls dead of a bullet wound.

> This is the description of a man who is acting in accordance with his inner volitions, a man whose body is unimpeded and unconstrained in its motions, these motions being the effects of those inner states. It is hardly the description of a free and responsible agent. It is the perfect description of a puppet.
>
> (1992, 46)

On the conditional-analysis view of compatibilism Fischer sketched above—but ultimately rejects (Fischer 2007, 50)—the agent here is free. If he wanted to raise his hand, he would. If he wanted to pull the trigger, he would. The problem, says the libertarian critic, is that it does not seem right to say that the agent is free in this case. The agent has not freely fired the gun. The ingenious physiologist has caused the relevant mental states and thereby the bodily actions that follow. Having been manipulated in this way, we ought not conclude that the agent is free.

The slippery slope has now been prepared. The agent in this example is not free, but there is no relevant difference between one person getting his desires/reasons from the physiologist's machine and another getting them from natural neurophysiological events. In both cases, the agent is passive, a slave to whatever internal mental states arise. In neither case should the agent's choices count as free. "The conclusion is that an agent is not morally responsible if he's causally determined to act by factors beyond his control even if he satisfies the compatibilist conditions" (Pereboom 2022, 34).

This is far from the final word on the matter. The back-and-forth between libertarians and compatibilists has continued for decades, with compatibilism being the majority view among philosophers (https://survey2020.philpeople.org/survey/results/4838). (One free-will skeptic from Chapter 1 is dismissive, considering it "verbal acrobatics" (Hossenfelder 2022, 131).) My goal in this chapter is not to settle the debate but rather to provide an overview of the terrain. Suffice it to say that if I can show that libertarianism survives the objections raised by physics, then compatibilism survives as well. (The compatibilist may deny needing any help in this regard, which is fine.) In fact, there may be a way of giving both sides what they want, which is what Chapter 7 aims to show. For now, let's press on to some distinctions that are now common in the free-will literature. While I treat them as types of libertarianism, there are compatibilist versions as well.

2.3 Leeway and Sourcehood

All libertarians agree that free will requires your choices to, in some way or other, be up to you. The forking paths analogy is one way of understanding "up to you." *You* choose which path becomes actual. This has traditionally been the central libertarian concern. Under determinism, there are no live options, and so all but one of the paths are merely conceptual. Let's say a bit more about the forking paths and then consider an alternative.

2.3.1 Leeway

The standard name for the forking paths approach is *leeway* freedom. It takes free will to mainly be about one's ability to do other than what one does. The leeway future is open, and the ways in which the future actually proceeds depend in part on our free choices. At the heart of leeway libertarianism is the Principle of Alternative Possibilities (PAP): "a person is morally responsible for what she has done only if she could have done otherwise" (Robb 2020). If Heather uses a hammer to smash the windshield of Chuck's car, Heather is morally responsible for the damage. She could have chosen not to do so. But if Karl and Chris—both competitive powerlifters—put the hammer in Heather's hands and physically force her to smash the windshield, she is not morally responsible. Heather had no choice; she could not have done otherwise. As Timpe points out, however, PAP is focused on moral responsibility rather than free will itself (2012, 70). If one prefers to keep moral responsibility out of the conversation for now, he proposes AP: "a person has free will only if he could have done otherwise," if he has the ability to do other than what he actually does (2012, 70). Leeway then amounts to something like the following. If ϕ is some action that an agent *S* chooses to do at some time *t*, then "agent *S* has the ability to choose or do otherwise than ϕ at time *t* if and only if it was possible, holding fixed everything up to *t*, that *S* choose or do otherwise than ϕ at *t*" (O'Connor and Franklin 2021, sec. 2.2). The "holding fixed" qualifier is important for libertarian freedom. Even a compatibilist can agree that if the antecedent conditions were different, so would be the outcome.

Leeway libertarianism is just the sort of freedom that determinism appears to be incompatible with. In a fully deterministic world, there is no right branching in time. It might seem as if there are different ways that events might go, but that is only because we are not Laplacian demons. As far as we know, a pair of dice might stop rolling in one of thirty-six different ways. A Laplacian demon would instead know the one and only "branch" that represents reality. The others are merely due to our ignorance of the physics involved. For things to have gone otherwise would require that the conditions had been different. Say that a golfer successfully hits a ball into the hole. The only way for the golf ball not to have gone in the hole is for it to have been struck differently, all else being equal. But under physical determinism, the motion of the golf club, the arms of the golfer, and indeed the intentions of the golfer are all determined by the laws of nature coupled with the initial conditions of the universe. If all events were the same up to the point where the golfer swings, then the ball must go in the hole, just as a Laplacian demon would have foreseen.

Leeway freedom seems clear enough and is likely what most people have in mind when it comes to their own free will. Why then have libertarians downplayed it in recent years? Many have been persuaded by "Frankfurt-style" counterexamples (Frankfurt 1969; Fischer 2010). Consider a case of electronic diet control. Say that a scientist invents a smartwatch that can detect one's food choices. The watch can accurately measure brain activity in such a way that it detects when you intend to make a healthy choice and when you intend to cheat on your diet. Let's say that you go into a food court capable of receiving signals from your smartwatch. Salad is on your diet plan. Pizza is not. If the watch detects that you intend to pick out a salad or some other healthy option, it does nothing. But if you intend to cheat on your diet and get pizza, the device activates, and all the pizzas are automatically lowered below the counter, and a "Sold Out, No Pizza Today" sign appears in their place. Now say you chose a salad. Should we praise you for sticking to your diet? It seems so. After all, that was in fact the choice you made. However, you could not have done otherwise. Had you intended to get pizza instead, that option would have been removed from the counter.

Would an agent have free will in such a scenario? Not if AP is a necessary condition. Given the responses built into the watch and food court, there is no scenario in which you are able to buy pizza. That "branch" may seem like a live option, but it is not. Some compatibilists applaud this as showing that the agent is free even when leeway branching is impossible. What the agent lacks, they say, is *libertarian* freedom. This has prompted a number of salvos back and forth about whether some form of libertarian freedom escapes the Frankfurt counterexamples (McKenna 1997; Kane 1998, chap. 4; Widerker 2011; O'Connor and Franklin 2021, sec. 2.3), leading some libertarians to no longer think that AP is the central concern:

> [The] exclusive focus on alternative possibilities in debates about compatibility [between determinism and free will] is a mistake. The fact that these debates have tended to stalemate over differing interpretations of *can*, *power*, *ability*, and "could have done otherwise" is a symptom of a deeper problem—namely, that AP alone provides too thin a basis on which to rest the case for incompatibilism: the Compatibility Question cannot be resolved by focusing on alternative possibilities alone. This does not mean that alternative possibilities have no role to play in free will debates, as some compatibilists would have us believe. But it does mean that their role is more complicated than is generally recognized.
>
> (Kane 2002, 407, quoted in Timpe 2012, 163)

Let's consider a more recent approach to libertarian freedom.

2.3.2 Sourcehood

Thus far we have assumed that the challenge determinism poses for free will is that it would block the conditions necessary for PAP and AP. Instead of an open future, there would only be one way that events can and will unfold. But this diagnosis of the tension presupposes leeway libertarianism. Source libertarians say the problem lies elsewhere.[13] The issue is about control, not the openness of the future:

> Frankfurt, Fischer, and company have gotten wrong what intuitively drives PAP. They think that the kind of control demanded by PAP is the kind of control in which we have it within our power to bring about some range of different kinds of alternatives—like jumping up into the air, or instead doing a little soft shoe; but what intuitively drives PAP is the kind of control needed in order for us to avoid being the author of a *particular* act and thus avoid being responsible for the production of *that* particular action… The issue for incompatibilists here is whether the will which displays these features places my stamp upon the world, and whether it is up to me or not to have that particular stamp or some other as my mark upon the world.
>
> (McKenna 1997, 73–75)

It is not that alternative possibilities are completely irrelevant to source libertarian worries about determinism. They just aren't the primary focus.

Let's change the diet-control Frankfurt case a bit to illustrate. Say that instead of the smartwatch sending a signal to the food court when it detects the wearer's intention to get pizza, the watch instead directly alters one's brain chemistry. If the watch detects a healthy choice, it does nothing. But if the watch detects an unhealthy choice, it beams a signal directly into the agent's brain that changes the desire to salad instead. While users cannot choose pizza in this scenario, that is not the reason we should say that they lack free will. Sourcehood libertarians instead argue that such agents would lack sufficient control over their own decisions. If the watch activates, it is the watch and its signal that are causally responsible for the "choice," not the rational deliberation of the person. In those cases, agents would not have libertarian freedom, but it is because of their lack of control in those cases, not the inability to do otherwise.

The same idea applies to physical determinism. Baseballs do not have free will, but not merely because they lack the ability to do otherwise. It is instead because they are wholly passive. The trajectory of the ball is determined by how it was hit, the wind conditions, and the laws of mechanics. The same basic picture applies to all material systems, including, says the determinist, human brains. The present state of a human brain and body is due to its previous state, sensory inputs, and the relevant laws of nature. What the agent perceives as desires, reasons, and intentions are effects, not causes. That, says the source libertarian, is the challenge of physical determinism. Persons would be unable to exercise any active control:

> Source Incompatibilism requires that the source of a free action cannot have sufficient causal antecedents that are external to the agent in question; it instead requires that those factors that are outside the agent's volitional structure are not sufficient for the agent's choosing in a particular way.
>
> (Timpe 2012, 142)

Freedom requires that the agent provides a necessary, proximate cause for the action. Libertarian incompatibilists believe the will fits that description. Determinist incompatibilists believe it does not.

> [The] core incompatibilist thought is that an agent is the ultimate cause of his action only if he contributes some necessary ingredient to it that cannot be traced back to causally sufficient conditions obtaining independently of his. In this way, only at indeterministic worlds, the incompatibilist can argue, does an agent contribute something distinctive or unique to the world, something that cannot be exhaustively traced to conditions outside of his own agency.
>
> (McKenna 2010, 437–38)

Once again, the challenge of determinism is that it would disallow one's causal contribution to events and thus prevent any sort of control over them.

If one believes in free will, says the incompatibilist, it must be libertarian freedom. Source libertarianism says that the agent's choices are both necessary for bringing about one's actions and subject to that agent's control.

We can now see more clearly how the skeptical arguments from Chapter 1 map onto this landscape. Carroll's "Core Theory" argument is a direct assault on leeway libertarianism, as it seeks to close off any physical space for alternative possibilities. Hossenfelder's aggressive reductionism threatens sourcehood libertarianism by insisting that any supposed "agent" is nothing more than a passive conduit for determined particle interactions.

2.4 The Question of Causes

"Bringing about" language is causal, but how should it be understood? The question opens onto a sprawling literature that is difficult to address. Among recent analyses of causation, many adopt metaphysically lean approaches; the most minimal of these are regularity theories, exemplified by Hume's account of causation as mere constant conjunction between events (Psillos 2010). At the other extreme are accounts that involve relations between universals (Tooley 1977) or dispositions/causal powers (Mumford 2004), the latter being the philosophical descendant of Aristotelian essences. O'Connor tries to narrow the field by arguing that the thinnest accounts of causation are inadequate for libertarian freedom (2000, 67–68). To support "bringing about" language, he says, one needs a view in which causes provide some sort of *oomph*—where *oomph* has nearly become a technical term in this discussion. Causes produce their effects. There are two main ways libertarians try to spell this out, but first let's acknowledge a contrarian view.

2.4.1 Noncausal Freedom

Not all libertarians agree that free actions are caused (Clarke 2003, 17–24). Noncausalists believe that while reasons can explain an agent's actions, reasons are not causes:

> Most of our actions we do for reasons. Most actions have explanations in terms of the agent's reasons for performing them... An important philosophical question about reasons explanations is whether the truth of such explanations requires a causal connection between the explaining factor (cited or implied) and the action.
>
> (Ginet 2008, 229)

Ginet's answer is that they do not. Reasons explain, but they do not cause.[14] It is the physical determinist who believes that all actions are caused. Noncausal theorists deny this. Causation is not what links agents to their actions.

Most source libertarians believe instead that the noncausal approach fails to provide sufficient control on the part of the agent, much like leeway accounts fail to do so. Libertarian freedom requires not only that determinism be false. What goes on in the mind of the agent must have a causal connection to the action; otherwise, the agent has not brought it about. And if the agent has not contributed to bringing the action about, the agent is not morally responsible for the action.

The next two positions agree that what free will brings about is an event, either a choice (mental state) or action (bodily movement) (Timpe 2012, 23). They differ over what causes these events.[15]

2.4.2 *Event Causation*

Normally, we think of events causing other events, as in "the impact of the rock causes the pond to ripple." The event-causal libertarian sees free choices within a sequence of mental states, where being in a particular mental state is itself an event. The sequence will include such things as desiring, intending, and having reasons to act. If it is raining, my windows are open, and I don't want my floors to get wet, then I have a reason for closing the windows, which I intend to do. Reasons here play a causal role, but a non-deterministic one. Randolph Clarke gives this example:

> Leo is deliberating about whether to play golf this afternoon, something that he very much enjoys doing. On the other hand, he wants to finish soon the drawings for a design project on which he is working. After briefly considering the matter, he decides to play golf. His having a desire to play golf and a belief that the design project can wait until tomorrow, together with his having the intention to make up his mind, nondeviantly cause his decision, but neither these events nor any other complete cause determines the decision. Until he makes his decision, there remains an objective chance that he will instead decide to work on the design project, this decision's being caused, in an appropriate way, by his having the desire to finish the drawings soon and his recognizing that finishing it soon precludes his playing golf...
>
> (2003, 30)

The notion of objective chance (or objective probability) is important here. What it does not refer to is a subjective degree of belief. If someone asks how likely I think it is that the Miami Dolphins will go to the Superbowl next year, I would respond, "I think there is perhaps a 30% chance of that happening." My interlocutor's own subjective probability about this might be higher or lower, and we would both be correct. That's because we are merely reporting how things seem to us rather than describing some objective, mind-independent state of affairs.

Objective probabilities are different. If I were asked how likely it is that a pair of dice will roll a 12, I would say 1/36. If someone disagrees, they are simply wrong. Given the number of sides and assuming the dice are fair, there are 36 equally possible outcomes. The right answer does not depend on how confident anyone is about the outcome.

While this example illustrates the difference between objective and subjective probabilities, it isn't quite right. Under physical determinism, the outcome of all such events is fixed by the laws of nature. The claim that Pr(rolling a 12)=1/36 is based on incomplete information. Unlike us, a Laplacian demon would know precisely whether the dice will roll a 12 or not. To such an intelligence in a fully deterministic world, the only objective probabilities are 1 and 0. A given event will either happen or it will not. If there are events having true, objective probabilities other than 1 and 0, then those events are not deterministic.

Let's go back to Clarke's example. When he says there is an objective chance (not equal to 0 or 1) that Leo will either go golfing or continue to work, he is denying that the choice is determined. No set of reasons guarantees that Leo will make a given choice. Both are live options, and a Laplacian demon would not be able to perfectly predict the outcome.[16] A given set of reasons merely raises the objective probability of a choice, but the choice itself is ultimately up to Leo. "It is said that he was able to do otherwise, that he had a choice about what he did, that what he did was up to him, that he determined, himself, what he did, that he was, in the relevant respect, an ultimate source, or origin, or initiator of his decision" (Clarke 2003, 30). Clarke adds that the ability to do otherwise here (leeway) is less important than that Leo exercises "active control" (2003, 30).

This leads to a well-known objection: the problem of luck (Mele 2006; Pereboom 2007, secs. 6–8; Timpe 2012, chap. 10). (There is a version of the problem for compatibilism as well. Also note that the problem here is of great interest to philosophers. Others might want to move along.) Consider Leo again. In the possible world that Clarke describes, Leo decides to play golf. But in another possible world identical to this one up to that point, Leo decides instead to be more responsible, work on his design project, and ensure that it is in on time. Since the possible worlds are identical up to that point, nothing in the first world accounts for why Leo chose to play golf, and nothing in the second accounts for why he chose to continue work. Leo's choice is indeterministic, a matter of objective probability.

Now the problem. If this description is correct, it appears that the decision is random. There is some objective probability of event *x* and another for event *y*, but which occurs is a matter of chance or luck. How is Leo intrinsically different from, say, the character Two-Face in the Batman comics, who makes decisions by flipping a coin? For Two-Face, the randomness lies in an external mechanism. For Leo, it is somehow built into the process for making the decision. Neither fits the libertarian description of a free choice. Neither allows one to "place his stamp" upon events, as McKenna put it. This is especially problematic if the key concern about free will is moral responsibility. Say that you are Leo's boss in the second possible world. Should you praise Leo for his diligence? Well, not if it was merely a matter of chance that he continued to work. Leo just got lucky. Both outcomes were live options up to that point, and which option he "chose" was a matter of indeterministic, objective chance. Dan Speak sums up the problem this way:

(1) Indeterminism just is (or entails) luck.
(2) Luck just is (or entails) lack of control.
(3) Control is necessary for freedom and responsibility.
(4) Therefore, indeterminism is incompatible with freedom and responsibility.

(Speak 2011, 33, quoted in Timpe 2012, 164–65)

This is the reason quantum mechanics provides limited help to the libertarian. Not only are indeterministic events not sufficient for free will, but indeterminism also leads directly to the problem of luck.

The literature at the end of this rabbit trail is substantial. Most libertarians deny that making a choice is merely a matter of objective chance or indeterminism. Establishing what it is, exactly, is harder, but most agree that agents have more control than how Leo is portrayed here.[17] "If there is no means by which I can take advantage of this looser [indeterministic] connectivity in the flow of events, its presence can't confer a greater kind of control, one that *inter alia* grounds moral responsibility for the action and its consequences" (O'Connor 2009a, 192). How then can such control be established? O'Connor and others do not believe that it will be found within some new version of event causation. An alternative approach is needed.

2.4.3 Agent Causation

The problem of luck raises the question of how agents influence their own mental states. According to some libertarians like O'Connor, there doesn't seem to be any space under event causation where the agent is clearly able to affect the sequence of events. But perhaps events are not only caused by other events. Perhaps actions and intentions are not caused by other mental states but by the agents themselves:

> On these agent-causal accounts, then, an agent is in a strict and literal sense an originator, an initiator, an ultimate source of her directly free action; she is an uncaused cause of that behavior, and one whose causing of that behavior is not causally determined.
>
> (Clarke 2003, 134)

While a person is causally influenced by many different factors, both rational and nonrational, these influences are not themselves sufficient for the agent's choices/actions. Tracing the causal chain back from an action leads to the agent, rather than mental states (like reasons) and circumstances outside of the agent's control. For O'Connor, what the agent principally causes are intentions:

> [Parallel] to event causes, the distinctive capacities of agent causes ("active powers") are grounded in a property or set of properties. So any agent having the relevant internal properties will *have it directly within his power to* cause any of a range of states of intention delimited by internal and external circumstances.
>
> (2000, 72)

The agent possesses the causal power to bring about free choices.

Agent causation seems to solve the question of control. Free will is not found within an indeterministic sequence of mental states. At some point, the agent himself enters the causal chain.[18] So why not embrace agent causation? There is one overarching concern raised by the free-will skeptics in Chapter 1: agent causation conflicts with physics. There are two ways such a conflict might occur. First, let's say that physical determinism is true. Pereboom picks up the objection here:

> On the route to action that results from an undetermined agent-caused action, changes in the physical world, for example, in the agent's brain or some other part of her body, result. But at this point, we would then expect to encounter divergences from the deterministic laws. This is because the alterations in the brain that result from the causally undetermined decision would themselves not be causally determined, and so they would not be governed by deterministic laws... Thus, it seems that agent-causal libertarianism is not reconcilable with the physical world's being governed by deterministic laws.
>
> (2022, 11)

This objection is, in a sense, old news insofar as all incompatibilists think that libertarian freedom conflicts with determinism. There is nothing special about agent-causation here. All libertarians must show how such freedom is possible in a world governed by natural laws.

Second and more important, agents are not the *kind* of causes recognized by science. While there are many different philosophical accounts of causation, they all involve events. In fact, agent-causal libertarians also typically believe that events cause events *except* in the case of agent causation. That's a problem. Special exemptions of this sort feel ad hoc, and agent-causal theorists know it. Making sense of libertarianism is not helped if it requires a sui generis form of causation.

In response to this, O'Connor and others have moved toward the broader notion of *substance causation*: substances cause events (O'Connor 2014). Substances are the sorts of objects one encounters on a daily basis, everything from rocks and chairs to cats and people (Koons and Pickavance 2017, 285). Some argue for a wholesale replacement of event causation with substance causation (Lowe 2013); others believe that different types of entities have different causal powers (Clarke, Capes, and Swenson 2021, sec. 3.3). For our purposes, the key here is that agents are just one type of substance, and so agent causation would no longer be sui generis.

But there is a cost. Substance causation generally fits within a dispositionalist/causal power metaphysic (O'Connor 2009a). On that view, entities have intrinsic dispositions to behave in various ways. Electrons, for example, do not move away from each other because it is a law of nature, but rather because they have the disposition to do so. They also have the disposition to be attracted to other bits of matter, which is ultimately what gravity *is* (Dorato and Esfeld 2016, 410). Causal powers/dispositions/capacities are responsible for the causal regularities in nature, rather than the laws of nature, as nomological realists hold. Metaphysicians tend to favor a causal powers view. Philosophers of science are often more skeptical (Smith 2002, 252).

There is, however, a less metaphysical way of thinking about this. As Jessica Wilson argues, there is nothing objectionable about "causal powers" if what one means is merely the causal contribution that a thing makes:

> [Talk] of powers is simply shorthand for talk of what causal contributions possession of a given feature makes (or can make, relative to the same laws of nature) to an entity's bringing about an effect, when in certain circumstances. That features are associated with actual or potential causal contributions ("powers") reflects the uncontroversial fact that what entities do (can do, relative to the same laws of nature) depends on how they are (what features they have).
>
> (2021, 32)

A causal power in this sense is roughly what an entity is able to do, the causal contribution it makes in some set of circumstances. One can remain agnostic about the underlying metaphysics—dispositions, laws of nature, or the denial of both (as Humeans prefer).[19] Everyone can use the language of "causal powers" in Wilson's thin sense without committing to any deeper metaphysics.[20] Wilson then shows how agent causation can be similarly understood:

> Concerns about agent causation often stem from concern about causes being substances rather than events. One might reasonably deny, however, that making sense of agent-causation requires thinking of agents as "substances" or positing an associated new form of causation, as opposed to thinking of agents as objects of a distinctive variety, no more problematic than the distinctive kinds of objects (cells, planets, plants, and so on) posited by the various special sciences, which are also supposed to be capable of causing various effects.
>
> (2021, 270)

The Sun causes the temperature on Earth to rise, and water causes salt to dissolve. Any entity can be treated as a substance in this sense. So yes, human beings are substances that have (thin) causal powers, and one of them is to be the source of free decisions, all of which can be affirmed without the baggage of substance causation:

> My own view is that there is no need to introduce a new variety of causation, much less a new substance, in order to accommodate the sort of nomologically transcendent freedom that agents possess (or should be thought to possess), on a libertarian view.
>
> (Wilson 2021, 267)

Causation is already vexing enough. Wilson believes that libertarians ought not invent another type of causation if they can help it.

But wait. There was, after all, a problem that agent-causal libertarians were trying to address: How is it that agents exercise control over events? Most substances, even in Wilson's thin sense, don't have free will. What is it about agents that gives them *this* causal power? The answer is not some new form of causation. It is instead a matter of emergence. "[Agential] control involves a fundamentally novel power to choose in a way that is not a matter of either deterministic or indeterministic nomological processes" (Wilson 2021, 269).[21] The emergence literature is where the analysis of such novel powers is found—something that O'Connor has been saying for some time (2009a, 191). This is why emergence will become our focus starting in Chapter 7. For now, I will adopt Wilson's thin sense of causal power and substance causation. We do not need a heavyweight understanding of either to talk about things—including agents—bringing about effects. How free agents come to have this capacity and how it amounts to control will have to wait for a bit more groundwork to be done.

2.5 Looking Ahead

This concludes an all too brief overview of an extensive debate. Still, there should be enough here for the uninitiated reader to get oriented. Let's make a few connections with what the free-will skeptics had to say in Chapter 1.

Which physics-based concern is most pressing often depends on whether one is a leeway or sourcehood libertarian. Consider physical determinism and the laws of nature. The laws are commonly thought to be all-encompassing. Every physical event is within their scope. If those laws are deterministic, then any sort of branching or forking paths in the future is merely conceptual. We might not know how events are going to unfold, but given the laws and the initial state of the universe, there is only one way they can go. Leeway libertarianism would therefore be impossible. This was the sort of thinking behind the Consequence Argument (Section 2.21.3) and the Laplacian Demon. Even if quantum mechanics undermines strict determinism, it is at best unclear whether the branching of indeterministic events has anything to do with free choice. The leeway approach alone does not show how the agent can exercise control to move from one branch to another.

The main challenge for sourcehood is reductionism. If the purported agents with free will are made of particles in the Standard Model, many physicists will insist that all the abilities and properties of such agents be compatible with the properties of those particles. But we know the laws in play in the Standard Model, and those laws have no room for agents to exercise such control. For sourcehood libertarianism "to exist, it would have to be possible for human beings to overcome the laws of physics just by thinking" (Carroll 2016, 381).

These are formidable challenges, and yet I believe they can be overcome. The first step is a closer examination of determinism.

Notes

1 As Taylor Cyr points out (private correspondence), while Pereboom is a free-will skeptic, he does not endorse determinism. To see how determinism and free-will skepticism might diverge, see Pereboom (2022, 38–39).

2 It is often claimed that open theists deny omniscience. This is false. Most believe that God knows everything there is to be known, but events in the future do not exist. God's not knowing the future is akin to God's not knowing the age of the current king of Michigan. There is no such being, and so this is not a matter of ignorance on God's part.

3 One proposal is that God merely foresees future events. God does not make those events turn out a particular way, at least those involving free decisions. Whether that helps will depend on a distinction to be made in Section 2.3.

4 This characterization relies on a necessitarian view of causation whereby causes bring about their effects in a non-probabilistic way. Causation, as we will see, is understood in a wide variety of ways.

5 This is only true for the most part. We will consider exceptions in Chapter 3.

6 The primary example was a version of Leibniz's Principle of Sufficient Reason, "Current events are connected with preceding ones by a tie based upon the

evident principle that a thing cannot come to existence without a cause which produces it" (Laplace [1814] 1902, 3). See van Strien (2014) for more.

7 If Copenhagen is indeterministic, but Bohm and Everett/DeWitt are deterministic, how can they all be empirically equivalent? Because deterministic systems can behave in random-looking ways. Chaos theory, which is the study of unpredictable systems, is based on deterministic equations, and relatively simple ones at that.

8 To summarize Fisher's critique, while Sapolsky provides impressive scientific detail about how our brains and behavior are shaped by genetics and environment, he never explains why this causal shaping should eliminate free will. His core fallacy is making a logical leap from specific cases where causal understanding does eliminate responsibility (like epilepsy and schizophrenia) to concluding that all causal explanations of behavior eliminate responsibility.

9 As Cyr points out (private correspondence), this way of characterizing free action is not limited to libertarians. This will become clearer in Section 2.3 when we distinguish the notions of leeway and sourcehood.

10 Determinists agree but see this as a reason to reject a realist interpretation of ethical obligation (Caruso 2021) or at least those rooted in the idea that one deserves praise/blame or reward/punishment for past behaviors (Pereboom 2007, sec. 10).

11 More precisely, the conflict is with leeway libertarianism, which will be discussed more fully in the next section.

12 Some libertarians describe this as the ability to "change the future," but that seems to presuppose that the future already has some way of being. I take the phrase to be shorthand for "change the future from what it would have been without some agent's action." For example, there is a picture on my wall that has been there for years. Unless someone acts, the future will include that picture hanging just where it is. The ability to change the future means that one could now choose to act so that events unfold differently.

13 There are also varieties of source compatibilism (Fischer and Ravizza 2000). On Fischer's version, also called "semicompatibilism," there are two types of control: guidance and regulative control. When driving a car, you guide the car by turning the steering wheel. You have guidance *control* when your actions behind the wheel are "based on your own reasons in the normal way" (Fischer 2007, 56). You also exercise regulative control in that you might have previously turned left instead of right—an example of AP. Fischer argues that guidance control is essential for free will, not regulative AP.

14 Also, see McCann (1998) and Ginet (2016).

15 This is where the distinctions cross-classify again. Compatibilists also believe in event causation, and some endorse agent causation, so these should not be understood as types of libertarian sourcehood (Timpe 2012, 24–25). The event/agent distinction is more central to libertarians, however, especially when it comes to the question of sufficient control.

16 Kane argues for exceptions where, for example, one's character has been so hardened that a given outcome is virtually certain. So long as it was the agent's free choices that led to the forming of his character, that person is still "ultimately responsible" (Kane 1998, 35).

17 Others argue that luck is not a single, unified problem but a family of issues (O'Connor forthcoming, 33; Timpe 2012, 163–64). Timpe believes this often leads to equivocation over "luck." This is not surprising, given that luck is related to chance and randomness, which also have many definitions across academic disciplines (Koperski 2022).

18 This line of thought also motivates agent-causal compatibilism (Cyr, private correspondence). See, for example, Markosian (2012).

19 For an introduction to Humeanism, dispositionalism, and nomological realism, see Koperski (2020, chap. 5).

20 Or as O'Connor puts it, the libertarian will be able to translate from one metaphysical scheme to the other. Several different approaches can support a minimalist understanding of libertarian free will (O'Connor forthcoming, 24).

21 Wilson cites Trenton Merricks as supporting a similar view:

> Sometimes my *deciding* to do such and such is what causes the atoms of my arm to move as they do. Presumably my so deciding won't ever be the *only* cause of their moving. There will also be a cause in terms of microphysics or microbiology, in terms of nerve impulses and the like. But at some point in tracing back the causal origin of my arm's moving (if it is intended), we will reach a cause that is *not* microphysical, that just is the agent's *deciding* to do something (Merricks 2001, 110).

Works Cited

Bohm, David. 1952. "A Suggested Interpretation of the Quantum Theory in Terms of 'Hidden' Variables." *Physical Review* 85 (2): 166–79. https://doi.org/10.1103/PhysRev.85.166.

Campbell, Joe. 2017. "The Consequence Argument." In *Routledge Companion to Free Will*, edited by Kevin Timpe, Meghan Griffith, and Neil Levy, 151–65. New York: Routledge.

Capes, Justin A. 2019. "What the Consequence Argument Is an Argument For: What the Consequence Argument Is an Argument For." *Thought: A Journal of Philosophy* 8 (1): 50–56. https://doi.org/10.1002/tht3.404.

Carroll, Sean. 2016. *The Big Picture: On the Origins of Life, Meaning, and the Universe Itself*. New York: Dutton.

Caruso, Gregg D. 2021. *Rejecting Retributivism: Free Will, Punishment, and Criminal Justice*. Law and the Cognitive Sciences. Cambridge: Cambridge University Press.

Clarke, Randolph. 2003. *Libertarian Accounts of Free Will*. Oxford: Oxford University Press.

Clarke, Randolph. 2011. "Alternatives for Libertarians." In *The Oxford Handbook of Free Will*, 2nd ed., 329–48. Oxford: Oxford University Press. https://doi.org/10.1093/oxfordhb/9780195399691.003.0016.

Clarke, Randolph, Justin Capes, and Philip Swenson. 2021. "Incompatibilist (Nondeterministic) Theories of Free Will." In *Stanford Encyclopedia of Philosophy*, edited by Edward N. Zalta. Stanford: Metaphysics Research Lab, Stanford University. https://plato.stanford.edu/archives/fall2021/entries/incompatibilism-theories/.

Dorato, Mauro, and Michael Esfeld. 2016. "The Metaphysics of Laws: Dispositionalism vs. Primitivism." In *Metaphysics in Contemporary Physics*, edited by Tomasz Bigaj and Christian Wüthrich, 403–24. Poznań Studies in the Philosophy of the Sciences and the Humanities 104. Boston: Brill.

Fergusson, David. 2019. *The Providence of God: A Polyphonic Approach*. Current Issues in Theology. Vol. 11. Cambridge: Cambridge University Press.

Fischer, John Martin. 2007. "Compatibilism." In *Four Views on Free Will*, edited by John Martin Fischer, 44–84. Great Debates in Philosophy. Chichester: Wiley-Blackwell.

Fischer, John Martin. 2010. "The Frankfurt Cases: The Moral of the Stories." *The Philosophical Review* 119 (3): 315–36.

Fischer, John Martin. 2023. Review of *Determined: A Science of Life without Free Will*, by Robert M. Sapolsky. *Notre Dame Philosophical Reviews*, November. https://ndpr.nd.edu/reviews/determined-a-science-of-life-without-free-will/.

Fischer, John Martin, and Mark Ravizza. 2000. *Responsibility and Control: A Theory of Moral Responsibility*. Cambridge Studies in Philosophy and Law. Cambridge: Cambridge University Press.

Frankfurt, Harry G. 1969. "Alternate Possibilities and Moral Responsibility." *Journal of Philosophy* 66: 828–39.

Gholipour, Bahar. 2019. "Does Free Will Exist? Neuroscience Can't Disprove It Yet." *The Atlantic*, September 10, 2019. www.theatlantic.com/health/archive/2019/09/free-will-bereitschaftspotential/597736/.

Ginet, Carl. 2008. "In Defense of a Non-Causal Account of Reasons Explanations." *The Journal of Ethics* 12 (3/4): 229–37.

Ginet, Carl. 2016. "Reasons Explanation: Further Defense of a Non-Causal Account." *The Journal of Ethics* 20 (1/3): 219–28.

Goldstein, Sheldon. 2017. "Bohmian Mechanics." In *Stanford Encyclopedia of Philosophy*, edited by Edward N. Zalta. Stanford: Metaphysics Research Lab, Stanford University. https://plato.stanford.edu/archives/sum2017/entries/qm-bohm/.

Hasker, William. 2011. "Divine Knowledge and Human Freedom." In *The Oxford Handbook of Free Will*, 2nd ed, 39–54. Oxford: Oxford University Press. https://doi.org/10.1093/oxfordhb/9780195399691.003.0002.

Hodgson, David. 2011. "Quantum Physics, Consciousness, and Free Will." In *The Oxford Handbook of Free Will*, edited by Robert Kane, 2nd ed, 57–83. Oxford Handbooks. Oxford: Oxford University Press.

Hossenfelder, Sabine. 2022. *Existential Physics: A Scientist's Guide to Life's Biggest Questions*. New York: Viking.

Kane, Robert. 1998. *The Significance of Free Will*. Oxfod: Oxford University Press.

Kane, Robert. 2002. "Some Neglected Pathways in the Free Will Labyrinth." In *The Oxford Handbook of Free Will*, edited by Robert Kane, 1st ed., 406–37. New York: Oxford University Press.

Koons, Robert C., and Timothy H. Pickavance. 2017. *The Atlas of Reality: A Comprehensive Guide to Metaphysics*. Chichester: Wiley Blackwell.

Koperski, Jeffrey. 2020. *Divine Action, Determinism, and the Laws of Nature*. New York: Routledge. https://doi.org/10.4324/9780429029110.

Koperski, Jeffrey. 2022. “The Many Faces of Randomness.” In *Abrahamic Reflections on Randomness and Providence*, edited by Kelly James Clark and Jeffrey Koperski, 13–27. Cham: Springer International Publishing. https://doi.org/10.1007/978-3-030-75797-7_2.

Laplace, Pierre Simon. (1814) 1902. *A Philosophical Essay on Probabilities*. Translated by F. W. Truscott and F. L. Emory. New York: Wiley.

Libet, Benjamin. 1985. “Unconscious Cerebral Initiative and the Role of Conscious Will in Voluntary Action.” *Behavioral and Brain Sciences* 8 (4): 529–39. https://doi.org/10.1017/S0140525X00044903.

Lowe, E. J. 2013. “Substance Causation, Powers, and Human Agency.” In *Mental Causation and Ontology*, edited by S. C. Gibb, E. J. Lowe, and R. D. Ingthorsson, 153–72. Oxford: Oxford University Press.

Markosian, Ned. 2012. “Agent Causation as the Solution to All the Compatibilist’s Problems.” *Philosophical Studies* 157 (3): 383–98.

McCann, Hugh. 1998. *The Works of Agency: On Human Action, Will, and Freedom*. Ithaca: Cornell University Press.

McKenna, Michael. 1997. “Alternative Possibilities and the Failure of the Counterexample Strategy.” *Journal of Social Philosophy* 28: 71–85.

McKenna, Michael. 2010. “Whose Argumentative Burden, Which Incompatibilist Arguments?—Getting the Dialectic Right.” *Australasian Journal of Philosophy* 88 (3): 429–43. https://doi.org/10.1080/00048400903233811.

McKenna, Michael, and D. Justin Coates. 2021. “Compatibilism.” In *Stanford Encyclopedia of Philosophy*, edited by Edward N. Zalta, 2021st ed. Stanford: Metaphysics Research Lab, Stanford University. https://plato.stanford.edu/archives/fall2021/entries/compatibilism/.

Mele, Alfred R. 2006. *Free Will and Luck*. Oxford: Oxford University Press.

Mele, Alfred R. 2009. *Effective Intentions: The Power of Conscious Will*. New York: Oxford University Press.

Merricks, Trenton. 2001. *Objects and Persons*. Oxford: Clarendon Press.

Mumford, Stephen. 2004. *Laws in Nature*. Routledge Studies in Twentieth Century Philosophy. Vol. 18. New York: Routledge.

Nahmias, Eddy. 2010. “Scientific Challenges to Free Will.” In *A Companion to the Philosophy of Action*, edited by Timothy O’Connor and Constantine Sandis, 345–56. Chichester: Wiley-Blackwell.

O’Connor, Timothy. 2000. *Persons and Causes: The Metaphysics of Free Will*. Oxford: Oxford University Press.

O’Connor, Timothy. 2009a. “Agent-Causal Power.” In *Dispositions and Causes*, edited by Toby Handfield, 189–214. Oxford: Clarendon Press.

O’Connor, Timothy. 2009b. “Conscious Willing and the Emerging Sciences of Brain and Behavior.” In *Downward Causation and the Neurobiology of Free Will*, edited by Nancey Murphy, George F. R. Ellis, and Timothy O’Connor, 173–86. Understanding Complex Systems. Berlin: Springer. https://doi.org/10.1007/978-3-642-03205-9_10.

O’Connor, Timothy. 2014. “Free Will and Metaphysics.” In *Libertarian Free Will: Contemporary Debates*, edited by David Palmer, 27–34. Oxford: Oxford University Press.

O'Connor, Timothy. forthcoming. "Probabilistic Explanation of Free Action."

O'Connor, Timothy, and Christopher Franklin. 2021. "Free Will." In *Stanford Encyclopedia of Philosophy*, edited by Edward N. Zalta. Stanford: Metaphysics Research Lab, Stanford University. https://plato.stanford.edu/archives/spr2021/entries/freewill/.

Pereboom, Derk. 2007. "Hard Incompatibilism." In *Four Views on Free Will*, edited by John Martin Fischer, 85–125. Great Debates in Philosophy. Oxford: Blackwell.

Pereboom, Derk. 2016. "Libertarianism and Theological Determinism." In *Free Will and Theism*, edited by Kevin Timpe and Daniel Speak, 112–31. Oxford: Oxford University Press. https://doi.org/10.1093/acprof:oso/9780198743958.003.0007.

Pereboom, Derk. 2022. *Free Will*. Elements in Philosophy of Mind. Cambridge: Cambridge University Press. https://doi.org/10.1017/9781108982511.

Psillos, Stathis. 2010. "Regularity Theories." In *The Oxford Handbook of Causation*, edited by Helen Beebee, Christopher Hitchcock, and Peter Menzies, 197–235.Oxford: Oxford University Press. http://oxfordhandbooks.com/view/10.1093/oxfordhb/9780199279739.001.0001/oxfordhb-9780199279739-e-0008.

Robb, David. 2020. "Moral Responsibility and the Principle of Alternative Possibilities." In *Stanford Encyclopedia of Philosophy*, edited by Edward N. Zalta. Stanford: Metaphysics Research Lab, Stanford University. https://plato.stanford.edu/archives/fall2020/entries/alternative-possibilities/.

Sapolsky, Robert M. 2023. *Determined: A Science of Life without Free Will*. New York: Penguin Press.

Saunders, Simon. 2010. "Many Worlds? An Introduction." In *Many Worlds?: Everett, Quantum Theory, and Reality*, edited by Simon Saunders, Jonathan Barrett, Adrian Kent, and David Wallace, 1–49. Oxford: Oxford University Press.

Schaff, Philip, ed. (1563) 1877. "The Heidelberg Catechism." In *The Creeds of Christendom*, vol. 3. New York: Harper & Brothers.

Shepherd, Joshua. 2017. "Neuroscientific Threats to Free Will." In *The Routledge Companion to Free Will*, edited by Meghan Griffith, Kevin Timpe, and Neil Levy, 423–33. New York: Routledge.

Smith, Sheldon R. 2002. "Violated Laws, *Ceteris Paribus* Clauses, and Capacities." *Synthese* 130 (2): 235–64.

Speak, Daniel. 2011. "Libertarianism, Luck, and Gift." *The Modern Schoolman* 88 (1–2): 29–49.

Taylor, Richard. 1992. *Metaphysics*. 4th ed. Foundations of Philosophy. Englewood Cliffs: Prentice Hall.

Timpe, Kevin. 2012. *Free Will: Sourcehood and Its Alternatives*. 2nd ed. New York: Continuum.

Tooley, Michael. 1977. "The Nature of Laws." *Canadian Journal of Philosophy* 7 (4): 667–98.

van Inwagen, Peter. 1989. "When Is the Will Free?" *Philosophical Perspectives* 3: 399–422. https://doi.org/10.2307/2214275.

van Strien, Marij. 2014. "On the Origins and Foundations of Laplacian Determinism." *Studies in History and Philosophy of Science Part A* 45: 24–31.

Wallace, David. 2012. *The Emergent Multiverse: Quantum Theory According to the Everett Interpretation*. Oxford: Oxford University Press.

Wegner, Daniel M. 2017. *The Illusion of Conscious Will*. New edition. Cambridge: MIT Press.

Widerker, David. 2011. "Frankfurt-Friendly Libertarianism." In *The Oxford Handbook of Free Will*, edited by Robert Kane, 2nd ed., 266–87. New York: Oxford University Press.

Wilson, Jessica M. 2021. *Metaphysical Emergence*. Oxford: Oxford University Press.

Zagzebski, Linda. 2021. "Foreknowledge and Free Will." In *The Stanford Encyclopedia of Philosophy*, edited by Edward N. Zalta. Stanford: Metaphysics Research Lab, Stanford University. https://plato.stanford.edu/archives/spr2021/entries/free-will-foreknowledge/.

3 The Challenge of Determinism

3.1 So What's the Problem?

Having defined the various positions in the free-will debate in Chapter 2, we now turn our attention to one of the primary concepts invoked by free-will skeptics: determinism. Arguments from physics often hinge on the idea that nature is deterministic, leaving no room for genuine choice. Let's start with a loose reconstruction of the free-will skeptics' main argument from Chapter 1.

> (1) Human beings are composed of particles (or some more fundamental stuff, like fields).

The precise ontology doesn't matter. The point is that we are made up of physical entities, whatever those entities ultimately turn out to be.

> (2) Bodies composed of those particles are governed by deterministic laws.

Hmm, is that right? What happened to quantum mechanics? Let's set that aside for now. Suffice it to say that at the relatively low energy states we occupy, the relevant laws are deterministic. This is why engineers learn so little about quantum mechanics. At the level of our experience, classical theories work well.

> (3) Determinism is incompatible with libertarian free will.

This was taken up in Chapter 2, although we were mainly concerned with understanding the different camps rather than what science says about them.

DOI: 10.4324/9781003681199-3

(4) Hence, libertarianism is false.

The free-will skeptics believe that science has settled the matter. The language of free will may be pragmatically useful but is strictly speaking false. Let's call this the Argument from Physics.

There are three critical concepts at work here. Premise (2) contains two of them. The first is *physical determinism*, which is the kind of determinism in view. The second is *law-governedness*, an early-modern innovation that says, in some sense, the laws of nature are running the show. The third is *composition* in (1). These are the subjects of this chapter, although one might wonder how much more there can be to say about determinism. It all seemed so straightforward in Chapter 2.

Philosophers of science tend to see determinism differently. In the first part of this chapter, we will consider what precisely physical determinism is and whether it was ever entailed by physics. Understanding how the idea has evolved will be helpful later in the chapter.

Even if one believes that quantum mechanics refutes determinism, libertarians often fail to appreciate the scope of natural law assumed in (2). Free-will skeptics believe that all events are governed by the laws, even if some of those laws are probabilistic, and that this fact is sufficient to rule out libertarianism. We will see why this makes quantum indeterminism largely irrelevant.

Finally, there is the question of composition. More precisely, what kinds of beings have the capacity to exercise free will? Once again, it's not as simple as it might seem. There are two main alternatives to the claim that we are nothing but atoms.[1] The first is well known but widely dismissed in scholarly circles: mind-body dualism. While the body is composed of atoms, says the dualist, there is an immaterial aspect of humans that is not. Premise (1), therefore, wrongly presupposes that physicalism is true. The second alternative is more esoteric: emergentism. The emergentist does not believe that every macroscopic entity can be wholly reduced to its constituent parts. In some sense, novel systems emerge from their constituent bases. What exactly "completely reduced" and "emerge" mean will be explored in later chapters. It is enough for now to know that emergentism accepts (1) but rejects the reductionist assumption that compositional truths tell the whole story.

The Consequence Argument (Section 2.2.3)[2] portrays physical determinism as something like the following. Take the overall state of the universe at an instant just after the Big Bang.[3] Every event after that is due to the laws of nature without exception. It's not that different from saying that the motion of the balls on a billiard table is due to the laws of mechanics and the state of the table just after the cue ball has been hit. The Laplacian demon is based on this same idea (Section 2.2.1.2).

The first important point of this chapter is that this picture lies somewhere between highly simplified and simply false. Physical determinism is not this straightforward, nor was it ever. If we are going to see how libertarian free will is possible, we first need a more precise understanding of physical determinism.

3.2 Determinism in Contemporary Physics

In the debate over free will, determinism is often thought of globally, as something true about the universe and its laws. In physics, determinism is local. It applies (or fails to apply) to individual systems.[4] Consider a simple, deterministic system: a block oscillating on an ideal spring—no gravity, no friction, and a perfectly elastic spring. Pull the block a bit and let it go. Determinism entails that if you put that same system in an identical environment and start it again from the same initial conditions, it would behave precisely the same way it did before. There is only one way the system can evolve from that state under the laws of nature. A similar principle holds for two identical systems.

So far I've left the notion of a *state* of a system unanalyzed. What exactly does that mean? The state is a complete description of a system's intrinsic properties at an instant in time. Different systems require different kinds of state descriptions.[5] Take the simplest: point particles. These are idealized bodies whose mass is concentrated in a mathematical point. The state of a particle can be fully specified by its position in space and its momentum.[6] To specify the state of a system of particles, one needs the position and momentum of each particle. This is the information a Laplacian demon would require, and it is vast. Six numbers are needed for *each* particle, three to specify its position in a three-dimensional space and three more for its momentum. All the possible states of a system of particles can be encoded in a mathematical space known as a "phase space" (from an older terminology in which each different state was called a "phase").[7] One point in a phase space represents the state of the entire system, no matter how many particles are involved.

This can be confusing. On the one hand, we're talking about point particles, meaning that those entities have no extension in our everyday three-dimensional space. On the other hand, we're talking about points in a phase space. These two senses of "point" and "space" should not be conflated. Point particles are idealized bodies with position and momentum. Points in phase space represent the *entire* state of a system in a mathematical space with many dimensions—six times as many dimensions as there are particles to consider. Most systems evolve over time in real space according to the relevant laws. This change is represented by a state point in phase space smoothly moving from one location to the next, carving a trajectory through the phase space. A phase space with some of these possible trajectories filled in is called a "phase portrait."

Let's apply this. When a system of particles operating under a given set of laws is deterministic, then from any given point in its phase space, a unique trajectory will be traced through that phase space.[8] A trajectory, again, represents how the system changes over time—from one state point to another. Identical systems in precisely the same state (i.e., having the same representative point in phase space) will evolve in the same way. What if the system were not deterministic? Then its trajectories in phase space would not be unique. They would at some point cross each other. A system at that crossing point might follow one trajectory or the other, and the laws would not determine which. While the notion of state gets more complicated for other types of systems, it is still the case that if a system is deterministic, then there is only one way it can evolve from any given state.

A phase space is a geometrical way of representing information encoded in a differential equation. Trajectories cannot cross in phase space if the underlying differential equation has a unique solution for each possible state. For one class of such equations, known as first order ordinary differential equations (ODE),[9] one can typically prove that unique solutions exist without actually having to derive them.

That's a lot of technical nuts and bolts, but it allows us to state Hossenfelder's objection from Chapter 1 more clearly:

> [All] laws of nature which we currently know work with those differential equations. These laws have the common property that if you have an initial condition at one moment in time, for example the exact details of the particles in your brain and all your brain's inputs, then you can calculate what happens at any other moment in time from those initial conditions. This means in a nutshell that the whole story of the universe in every single detail was determined already at the big bang. We are just watching it play out.
>
> These deterministic laws of nature apply to you and your brain because you are made of particles, and what happens with you is a consequence of what happens with those particles.
>
> (Hossenfelder 2020)

Differential equations are deterministic when their solutions exist and are unique. The trajectories in their corresponding phase spaces never cross, which means there is only one way the system can evolve from any given state. Since those equations accurately describe the physics of one's constituent parts, determinism is entailed by the physics of material bodies, or so the argument goes.

Hossenfelder overstates her case in several ways. First, not all systems can be modeled using first order differential equations and phase spaces. Those existence and uniqueness theorems for ODEs that prove determinism are not valid across the whole of physics, let alone chemistry or biology. Second, the theorems do not guarantee that anyone can *calculate* these solutions. The solutions exist within the realm of abstract mathematics, but knowing that they exist does not mean we can derive them. Third and most important, all this displays a common but naïve understanding of determinism in physics grounded, I believe, in a particular view of the history of physics. Most contemporary physicists will tell you that nature appeared to be fully deterministic until the rise of quantum mechanics proved otherwise. Let's see how their views about physics before the twentieth century tend to come about.

3.3 Laplacian Determinism

In Chapter 2, we touched on the most vivid illustration of physical determinism in the era of classical mechanics: the Laplacian demon (Section 2.1.1.2). Its computational abilities meant that "nothing would be uncertain and the future, as the past, would be present in its eyes" ([1814] f, 4). As the story is often told, Laplace was simply applying the known principles of physics at the time. The Laplacian demon worked as a thought experiment because physics had proved that determinism was true.

While this is a lovely bit of scientific folklore, it is also false. To understand what physics had and had not shown, let's first look back a few decades earlier.

The idea that the laws of nature govern all physical change goes back at least as far as Spinoza. "Nothing happens in nature that does not follow from her laws" (*Theologico-Political Treatise*, 6/19). Real contingency or randomness was forbidden in such a world (*Ethics*, Ip29). Leibniz, who contributed far more to early modern science than Spinoza, took a somewhat different route. There is his Principle of Sufficient Reason: each contingent event has a reason for why it occurs rather than some alternative. "[We] hold that there can be no fact real or existing, no statement true, unless there be a sufficient reason, why it should be so and not otherwise, although these reasons usually cannot be known by us" (Leibniz [1714] 1899, secs. 31–32). Leibniz also introduced the less well-known Principle of Continuity. Reduced to a slogan, it says "nature makes no leaps" (Leibniz [1765] 1916, 50). Any system in a given state S that evolves into some later state S' must smoothly proceed through all intermediate states. In more modern terms, this means that the trajectories in a phase portrait can have no gaps. State points move in a continuous fashion. For a system in any state, says Leibniz, there is a complete explanation for why it is in that state (Principle of Sufficient Reason), and part of that reason lies in the previous state from which it smoothly evolved (Principle of Continuity).

Laplace took the Principle of Sufficient Reason and gave it a distinctly causal interpretation. Each event has a cause:

> Present events are connected with preceding ones by a tie based upon the evident principle that a thing cannot occur without a cause which produces it. This axiom, known by the name of *the principle of sufficient reason*, extends even to actions which are considered indifferent … .
>
> (Laplace [1814] 1902, 3)

As van Strien shows, Laplacian determinism emerges from this doctrine plus the Principle of Continuity (2014a, sec. 5). Every event has a sufficient cause, and all causal change smoothly evolves from one event to the next. These principles provide the foundation for the Laplacian demon.[10]

The neglected point in contemporary discussions is that Laplace's thought experiment was not grounded in classical physics but in early modern metaphysics. "This is a completely general philosophical argument, and the whole story of the background of Laplace's determinism can thus be told without referring to mathematics or the equations of mechanics" (van Strien 2014a, 30). In other words, the physics was largely irrelevant; for Laplace, determinism was an established *metaphysical* doctrine. Telling the story that way is less compelling in our pro-science day and age than saying Laplacian determinism was entailed by Newtonian physics.

Perhaps Laplace has received too much attention. He is, after all, just one figure, and the Laplacian demon is merely one example in the broad sweep of science. One might grant that modern science evolved from a mishmash of metaphysics, alchemy, and other influences. Instead of focusing on the genesis of physics, let's consider classical mechanics once it was fully developed. If there is any basis for physical determinism, surely it must be there. Right?

3.4 Determinism in Classical Mechanics

The short answer is "mostly." Many classical systems meet the criteria spelled out in Section 3.1. They are modeled by differential equations, and those equations have unique solutions. Contrary to what one would expect given Hossenfelder's description, however, there are exceptions (Earman 1986, chap. 3; Earman 2004, sec. 3). I have discussed several of these elsewhere (Koperski 2020, chap. 6), but let's consider some key examples.

The Euler strut is a thin strip of metal attached to a base. Think of Figure 3.1 as looking along the thin edge of the strip so that its width extends into the page. Let's say that the strut is perfectly uniform so that no section is stronger or weaker than another. Now start loading the top with weights so that their center of gravity is directly over the strut—no extra weight to either the left or right. Eventually, the strut will reach a state of structural instability. Any further loading will cause it to buckle, bending one way or the other. Given its symmetry and the way the load is applied, no matter which way the strut buckles, it could have gone the other way. Nothing about the physics favors one outcome over the other. The buckling is indeterministic. This is one of many examples where indeterminism arises from some sort of symmetry in the system (Wilson 1989).[11]

Norton's Dome falls into a different class of examples (Norton 2008; Fletcher 2012). Consider a point particle at the very top of a frictionless dome but under the influence of gravity. This looks like a standard mechanical system, one that might even be suitable for textbooks. If the dome's surface has the proper shape, there is a class of solutions in which the

Figure 3.1 Euler Strut.

particle behaves just as one would imagine: it stays right where it is unless it is perturbed in some way. But there is another set of solutions in which the particle can start sliding off the dome at some arbitrary time without being pushed, which is decidedly not what is taught in freshman physics. The motion is indeterministic. A Laplacian demon would not be able to predict when it would occur.

Why is Norton's Dome such an unfamiliar case? Mainly because of principles that have already ruled it out as physically impossible. The equations violate a condition known as *Lipschitz continuity* that circumscribes mathematical descriptions deemed physically meaningful. Note that this restriction is not found within the laws of motion themselves. Lipschitz continuity is a metaphysical judgment about what the laws permit regardless of what the mathematics entails, judging, in effect, that "nature isn't like *that* even though the laws of mechanics allow it." Of course, no physicist would say he/she is doing metaphysics when appealing to Lipschitz continuity. Still, if we are talking about what is physically possible over and above what the laws of nature allow, that certainly looks like metaphysics.

While many scientists would object to this, there's no getting around metaphysics when doing physics. It's usually just buried so deeply in one's background assumptions and so widely shared across the scientific community that it is easy to miss. Not so when the issue is determinism, as the history of science continues to make clear.

To see why, let's start with an unappreciated fact: Physics has never operated under a single ontology. There were three broad approaches to classical mechanics, each with its own conception of matter (Wilson 2016; van Strien 2021, 3). The first is the most familiar. Laplace, Roger Boscovich, and Emil Du Bois-Reymond championed the idea of point particles—essentially atoms reduced to mathematical points so as not to have extension.

The second, often called "analytical mechanics," considers rigid bodies under constraint. Think of a bead sliding down a wire or a ball rolling along a track. The bead and ball are macroscopic objects treated as if they have no internal "give" or elasticity. The wire and the track limit their respective behaviors and the forces one must try to describe mathematically. Pioneers such as Jean le Rond d'Alembert ([1743] 2009), Fraser (1985), and Joseph-Louis Lagrange ([1811] 1997) showed how such an approach could solve problems that were hopelessly complex from a particle point of view.

Finally, classical continuum mechanics treats matter as having no gaps at any scale (Wilson 2016, sec. 6). Rather than discrete particles, matter is instead smooshed out in all directions. Think of the way water feels flowing over your hand. It doesn't feel like lots of tiny particles bouncing off your skin. We think of matter as molecular only because scientists tell us it is (for good reason). Before the existence of atoms was proven in the early twentieth century, continuum mechanics was not only useful but was a live option for modeling the true nature of matter.

None of these three was able to capture the whole of classical mechanics. Each relies on its own laws, principles, and conceptions of matter. Nor can they be harmonized. The "three basic approaches are not, from a foundational point of view, strictly compatible with one another: the modes of description favored as basic in one approach can only be treated as convenient approximations within the others" (Wilson 2009, 177).

Why this digression? Because the history of physical determinism follows a similar trajectory. Let's consider point particles again. Astronomers work at scales that allow planets and stars to be treated as individual points. While that may sound extreme, this approach allowed physicists to predict the discovery of Neptune. The orbit of Uranus seemed to violate Newtonian mechanics until they realized that it would be just what one would observe if another planet farther out were perturbing it. When astronomers were told where to point their telescopes, they found Neptune, a planet that was in a sense discovered on paper. Even before that discovery, the success of celestial mechanics led Laplace to believe that all bodies could be modeled the same way:

> The regularity astronomy shows us in the movements of the comets doubtless exists also in all phenomena.
>
> The curve described by a simple molecule of air or vapor is regulated just as certain as the planetary orbits; the only difference between them is that which comes from our ignorance.
>
> (Laplace [1814] 1902, 6)

So while we lack the computational capacity to predict the behavior of air molecules, the governing principles are no different than those used to predict the motions of planets.

We can now make one last observation about Hossenfelder's claims (Section 3.2). Point particles subject to Newtonian laws can be accurately described using ODEs. As we've seen, that class of equations is special because under standard regularity conditions (e.g., Lipschitz continuity), they have unique solutions. This means that for any given state of the system being modeled, there is only one way it can evolve over time, and this is sufficient to say that the system is deterministic. It is no coincidence, then, that Hossenfelder talks about particles. She is cherry-picking examples that most readily support her claim. For her conclusion to apply globally, one would have to show that all physical systems can be treated as collections of point particles.

The problem is that most real-world systems cannot be modeled this way, something that has been known for two hundred years (Fox 1974, 109–27). Not even common systems like water in a channel or a bending beam can be accurately modeled as collections of point particles. Such an ontology is inconsistent with wide swaths of science, including Fresnel's wave theory of light, Dalton's atomic theory of chemistry, and Ampère's electromagnetism. But even if that were not the case, determinism still fails in a point particle world. If such particles were to collide, the event would produce a singularity.[12] The laws of mechanics cannot say how the particles would behave immediately after the collision.[13]

Fortunately, the success of classical mechanics did not require point particles. Analytical mechanics instead deals with macroscopic bodies, often treating them as fully rigid and undeformable. The motion of such bodies is often constrained in some way, like a bowling ball rolling down a gutter. Rather than Newtonian vectors and forces in Euclidean space, analytic mechanics relies heavily on energy and *variational principles* that (usually) find the smallest value for some mathematical expression (Marion and Thornton 1988, 190–94). The power of this approach and how it differs from standard Newtonian mechanics is underappreciated. Nonetheless, it does not provide a safe harbor for determinism. Some collisions once again lead to singularities (van Strien 2021, 8). One example is a perfectly hard ball dropping from a chute centered directly over a knife edge. Some prominent physicists believed these singular points might be related to free will (van Strien 2014b, 175–76). Here is an excerpt from an 1873 letter by James Clerk Maxwell:

> Much light may be thrown on some of these questions by the consideration of stability and instability. When the state of things is such that an infinitely small variation of the present state will alter only by an infinitely small quantity the state at some future time, the condition of the system, whether at rest or in motion, is said to be stable; but when an infinitely small variation in the present state may bring about a finite difference in the state of the system in a finite time, the condition of the system is said to be unstable.
>
> It is manifest that the existence of unstable conditions renders impossible the prediction of future events, if our knowledge of the present state is only approximate, and not accurate.
>
> It has been well pointed out by Professor Balfour Stewart that physical stability is the characteristic of those systems from the contemplation of which determinists draw their arguments, and physical instability that of those living bodies, and moral instability that of those developable souls, which furnish to consciousness the conviction of free will.
>
> (Campbell and Garnett 1882, 211)

Maxwell thought that physicists were ignoring such cases and that paying more attention would "tend to remove that prejudice in favour of determinism" (Campbell and Garnett 1882, 213). In any case, there was no controversy over whether singular points are indeterministic.

Continuum mechanics overcame some of these problems but at a cost. Point particles and rigid bodies are mathematically tractable because they have limited degrees of freedom—a relatively small number of ways they could undergo change. If we take a rigid rod, for example, it can move along three different axes in Euclidean space and rotate along each. In a continuum system, there are infinite degrees of freedom due to matter existing at each location inside a body. Physicists employed idealizations to tame this complexity, like assuming that bodies and fluids are homogeneous as well as continuous.

As we've seen, collisions proved to be a problem for point particles and rigid bodies. This was not the case for continuum mechanics, which recognizes that no extended body is perfectly rigid. When objects collide and rebound, they deform. But describing this deformation requires replacing ordinary differential equations with partial differential equations. While this sounds like a trivial difference, the degree of complexity involved is significant. Unlike ODEs, we typically cannot prove that partial differential equations have unique solutions. Recall that existence and uniqueness formed the basis for believing that systems modeled by ODEs are deterministic—a crucial link in Hossenfelder's argument. Without this guarantee, we cannot say that continuum and fluid systems are globally deterministic, at least not based on physics alone. (Like Laplace, one could always use metaphysics to fill in the gaps.)

In short, there was never a way to do classical mechanics that was free of indeterminism. Still, many remained confident that nature was, in fact, fully deterministic, even if physics could not prove it to be so. As van Strien concludes, "many physicists seem to have taken for granted that all physical processes are uniquely determined, often without explicitly arguing for this, and without necessarily being committed to the idea that current theories of physics could yield a deterministic description of all natural processes" (2021, 10). Rather than an inference derived from physics, determinism was instead presupposed as a methodological principle. If current theories did not entail determinism, physicists assumed that we have simply failed to discover it. Prominent turn-of-the-century physicists made this clear:[14]

> The correctness of the position of "*determinism*" or "*indeterminism*" cannot be proven. It could only be decided if science were completed or demonstrably impossible. We are dealing here with presuppositions that one brings to the consideration of things... But during research, every thinker is necessarily a theoretical determinist.
>
> (Mach 1905, 277)

> A precondition for all scientific knowledge is the principle of unique determination of natural processes; in the case of mechanics, the unique determination of all motions.
>
> (Boltzmann 1905, 276)

> Science is deterministic; it is deterministic *a priori*; it postulates determinism, because without it, science could not exist.
>
> (Poincaré [1913] 1917, 244)

This is a somewhat different attitude than we have been led to expect. Even at the pinnacle of the classical era, determinism was a premise for doing science, not a conclusion. It was aspirational, a useful heuristic, not something derived from physics. Even today, the only way for a modern textbook on classical mechanics to ensure that determinism holds is to stipulate it (Arnold 1988, 2). This stands in stark contrast to the contemporary skeptics, who present physical determinism not as a methodological starting point for research but as a metaphysical conclusion delivered by it. This historical shift from philosophical premise to purported scientific fact is central to the confusion this book seeks to untangle.

While this is known to philosophers of physics, it is not well recognized outside of philosophical circles. My sense is that when philosophers talk about physical determinism, most think that classical mechanics underwrites their intuitions. They believe that until the rise of quantum mechanics, physics entailed determinism.

One might think physicists should know better, but again, my experience is that they are usually surprised when challenged about classical mechanics and determinism. Why is that? The main reason is pedagogical. They learn the mathematics of classical mechanics on the way to more interesting topics in quantum mechanics, particle physics, cosmology, etc. An oversimplified view of classical physics is just a stepping stone.

The main point of this section is that the universal, fully locked-down sort of physical determinism that Spinoza, Leibniz, and Laplace had in mind never materialized in science. This is the unappreciated flaw in how some determinists seek to deploy the Consequence Argument (Section 2.2.3).[15] It is not the case that every event in the past is a consequence of the laws of nature combined with the universe's initial state. Or at the very least if it is, physics has never shown it to be so.

3.5 Quantum Mechanics and Indeterminism

Some readers will wonder why I have taken this detour into the history of science. Hasn't quantum mechanics answered the question of determinism once and for all? Unlike classical physics, quantum mechanics is widely believed to be indeterministic. Some prefer the term "ontologically random" to emphasize that what is going on isn't just a lack of knowledge on our part. To understand the difference, consider dice. Their purpose is to add a measure of chance to games. But if their linear and angular momenta, the table's coefficient of friction, and other physically relevant information were known, how the dice will land could be calculated. There is a fact of the matter about the outcome, even though the players are not in a position to know the parameters that determine it. Quantum mechanics is different. Given its intrinsic indeterminism, not even a Laplacian demon could predict how such systems will evolve.

Libertarians typically take this to be good news. The stranglehold of physical determinism has been broken, and science has proved it! Kane describes what the consequences of this might look like:

> [If] any libertarian theory of free will is to succeed there must be some genuine indeterminism in nature to make room for it… This is the point … where some scientists want to bring modern quantum physics into the picture to help account for free will. Suppose there were quantum jumps or other undetermined quantum events occurring in the brain. We know that information processing in the brain takes place through the firing of individual neurons or nerve cells in complex patterns. Individual firings of neurons in turn involve the transmission of chemical ions across neuronal cell walls, stimulated by various chemicals, called neurotransmitters, and by electrical stimuli coming from other neurons. Some neuroscientists have suggested that quantum indeterminacies in the transmission of these chemical ions across the cell walls of neurons might make the exact timing of the firings of individual neurons uncertain, thus introducing indeterminism into the activity of the brain and making "room" for free will.
>
> (Kane 2005, 133)

Quantum mechanics is therefore potentially helpful in the fight against determinism and compatibilism.

This hope is not shared by most libertarian philosophers, however, or at least it is not thought to be highly significant. There are four reasons for this.

First, even if physics is ultimately indeterministic, that is far from sufficient for libertarian freedom. "Ontologically random" is not the same as "free." Hossenfelder presses this point:

> In quantum mechanics some events are truly random and cannot be predicted. Does this mean that quantum mechanics is where you can find free will? Sorry, but no, this makes no sense. These random events

> in quantum mechanics are not influenced by you, regardless of exactly what you mean by "you", because they are not influenced by anything. That's the whole point of saying they are fundamentally random. Nothing determines their outcome. There is no "will" in this. Not yours and not anybody else's.
>
> (Hossenfelder 2020)

Most libertarians recognize that free will requires some degree of control, but an ontologically random world allows for no more control than a deterministic one.

Second, even if there are indeterministic events at quantum scales, they seldom make their way into the macroscopic realm (Bishop 2023, 131–34; Honderich 1993, 65–67). Quantum effects typically only become observable in a laboratory setting, as Pereboom explains:

> For alternative possibilities to be significantly probable, there would have to be mechanisms that facilitate the "percolating up" of significant microlevel indeterminacies to the neural level, on the analogy of a Geiger counter that senses microlevel events and registers them at the level of the moving of a macrolevel indicator.
>
> (2022, 8)

The problem is that what happens at the quantum level tends to stay at the quantum level. The world of our experience looks classical, which is why so many of our intuitions are roughly Newtonian. This explains why students are hesitant to accept the counterintuitive descriptions of reality presented by relativity and quantum mechanics as accurate representations of our universe. For quantum indeterminism to be of help to the libertarian, it would have to manifest itself at the scale of brains (Sapolsky 2023, 217–22).[16]

Third, whether one takes quantum mechanics to be deterministic or not depends on which interpretation one favors. Introductions to quantum theory often rely on the (so-called) Copenhagen interpretation, in which some events *are* indeterministic. Many are surprised to learn that few physicists or philosophers of physics still accept that interpretation. Textbooks might still refer to "measurements" and "the collapse of the wave function" to help get the key concepts up and running (Becker 2018, 17–19), but most experts do not understand them the way Niels Bohr and others proposed.

One current favorite for replacing the Copenhagen interpretation is somewhat of a surprise. The Everettian or Many-Worlds view, once considered a novelty, has recently undergone a revival (Wallace 2012; Carroll 2019). To illustrate, consider the famous thought experiment known as Schrödinger's cat. A closed room is configured to produce a 50-50 chance of finding the cat dead or alive, where the outcome depends on random quantum events. Unlike the Copenhagen interpretation, in which one or the other randomly occurs, in the Many-Worlds interpretation, both outcomes are realized. The live cat is found in one world and the dead cat in another. For each event that the Copenhagen interpretation considers ontologically random, every possible outcome instead happens in its own world. A "world" here is more precisely an independent branch of the wavefunction for the whole of physical reality. Wavefunctions are typically used in quantum mechanics to describe the state of individual quantum systems, but the Many-Worlds interpretation takes a single, ever-branching wavefunction to be what physical reality ultimately is. To its inhabitants, each branch is essentially its own universe. For our purposes, what matters here is that the evolution of this wave function is deterministic. There is no ontological randomness.

Another well-known view is the de Broglie-Bohm interpretation, pioneered by Louis de Broglie and rediscovered by David Bohm (1952) (Goldstein 2017, sec. 3). Bohm's theory introduced a new unobservable entity, the pilot wave, which governs the motion of particles. While the behavior of the particles themselves is sometimes unpredictable and indeterministic, a particles-and-pilot-wave system evolves deterministically.[17] What looks like randomness in the particles is due to our incomplete information about how the pilot wave is influencing their behavior. Pilot waves are what Einstein called a "hidden variable." If we had access to all the underlying physics, we would see that quantum systems are deterministic.

My personal sense is that both physicists and philosophers of physics are trending toward views that deny ontological randomness and restore determinism in one way or another. If one of these deterministic views is correct, it would eliminate any ontological randomness that libertarians might try to make use of.

Fourth, some argue that determinism gets too much attention in this discussion. Recall premise (2) in the Argument from Physics:

> (2) Bodies composed of those particles are governed by deterministic laws.

This chapter has focused on determinism because it is one of the two incompatibilist positions in the free-will debate. As Michael Esfeld argues, however, it is not the determinism mentioned in (2) that is a threat to libertarian freedom but rather law-governedness. Regardless of whether deterministic or probabilistic laws govern nature, all events fall within their scope, and humans lack the ability to change them:

> If the decisions of human beings concerning the motions of their bodies influence neither the initial state of the universe nor the deterministic laws of classical mechanics, they do not influence the objective probabilities implemented in a fundamental stochastic law and an initial wave-function either...
>
> (Esfeld 2020, 93)

Objective probabilities were introduced in Section 2.4.2. The idea is that (at least some) probabilities capture real chance events in nature rather than subjective degrees of confidence a person might have about some claim. If one believes the probability of randomly drawing an ace from a deck of cards is 3/52, that person is wrong. The belief does not correspond to the objective, mind-independent probability (4/52). The probability of rolling a 9 with a fair dodecahedron die is 1/12, which in this case is grounded in the symmetry of the die. According to the Law of Large Numbers, the observed frequency of each face of the die should converge to 1/12 with a sufficiently large sample size. While dice and cards are simple examples, objective probabilities are often taken to be ubiquitous in nature.

What if free-will decisions were an exception? What if they do not conform to objective probabilities? "[Then] the stochastic law would be refuted, as a deterministic law would be refuted in case persons do not make the movements that are fixed by the law and the initial conditions" (Esfeld 2020, 93). Indeterministic laws are still *laws,* argues Esfeld, and they dictate the objective probabilities in nature. If free choices were to violate those probabilities, they would violate the laws of nature. To take a fanciful example, say that leprechauns supernaturally influence the behavior of the dodecahedron die from the previous paragraph so that it favors even numbers. That would mean that its long-term behavior did not conform to the law of large numbers. The objective, naturalistic probabilities would therefore have been violated. This, says Esfeld, is the real problem the laws pose for libertarianism. "What is at issue here is not determinism, but universal laws of nature, be they deterministic or not. If they are probabilistic, then the objective probabilities for our acts are the consequences of the laws of nature and events in the remote past" (Esfeld 2020, 96).

What drives some libertarians to a sourcehood rather than a leeway account of freedom (Section 2.3) is the issue of control. They say that free will is not about the ability to do otherwise but rather the control one has over one's actions. *I* choose to reach for my coffee mug. Nothing makes me do it. The challenge of physical determinism is that this decision would be the product of the laws of nature rather than the agent. Esfeld's point is that indeterminism is no better. If nature is law-governed, it doesn't matter whether the laws are deterministic. Suppose the act of reaching for my coffee mug violates a probabilistic law. In that case, Carroll's complaint is still correct: "In order for libertarian freedom to exist, it would have to be possible for human beings to overcome the laws of physics just by thinking" (Carroll 2016, 381). The standard distinctions in the free-will literature invite us to see determinism as the main threat to libertarian freedom. But "determinism" is often short for "deterministic control," and it's the control by the laws of nature that is at issue. Every physical event in the universe falls within their scope, regardless of whether those laws are indeterministic.[18]

The upshot of these four objections is that quantum mechanics is far less useful to libertarianism than often assumed in more popular sources. In the end, quantum indeterminism might not exist, and even if it does, that indeterminism is not found at the level of macroscopic organisms.

This chapter has shown that the question of physical determinism is more complex than physicists usually assume. Classical mechanics was never as deterministic as we tell our students. The arguments for physical determinism, while historically influential, have been considerably overstated, falling well short of the decisive, settled matter presented in Chapter 1. Still, the goal here is not merely to show that things are more complicated than the experts sometimes let on. To successfully defend libertarianism, I propose making the burden as great as possible. Let's assume that fundamental physics is fully deterministic in something like the Laplacian sense. Where in such a universe would there be room for libertarian freedom?

3.6 Two Ways Out

There are two approaches we will consider.[19] Both grant that physical determinism rules in a world of entities composed of subatomic particles (or fields—the exact ontology doesn't matter). What they deny is reductive physicalism, the claim that everything that exists can be reduced—in some sense of "reduced"[20]—to fundamental physics.

The first option, mind-body dualism, holds that there is an immaterial aspect of persons, often called a "soul" or "mind." The more familiar variety is *substance* dualism, famously traced back to Descartes. Mental states such as deliberation and sensory experience exist in the mind, says the dualist, not in the brain. Minds also give persons the ability to make choices. It is this substance in which free will "happens." For our purposes, the mind's key attribute is what it is not. It is not a physical thing.

Dualism avoids physical determinism in a straightforward way: The laws of nature do not apply to non-physical entities. While the laws of nature might be deterministic (as we've assumed), those laws are not the only influence on human beings. If human persons are not wholly composed of material stuff, premise (1) in the Argument from Physics fails to tell the whole story. And if free will is a capacity of immaterial minds, then physics can tell us little, if anything, about it.

The idea that minds are not bound by the laws of nature was at least suggested by Descartes[21] and made more explicit by Isaac Newton, John Locke, Samuel Clarke, mathematician Willem 's Gravesande, and natural philosopher Émilie du Châtelet (Pitts 2019, secs. 3.1–3.3). Physician Georg Ernst Stahl developed it further in the early eighteenth century (Ahnert 2014, 311), as did Leonard Euler, the preeminent mathematician-physicist of the mid-1700s (Breidert 2008, 103–4; Pitts 2019, sec. 6). As Stewart Goetz points out, belief in dualism was not originally motivated by determinism (2017, 523). The idea goes back at least as far as Plato, which predates any of our present concerns. In any case, dualism provides libertarians with an answer to the problem of determinism, one that many dualists are happy to acknowledge.[22]

The second option is emergence, which agrees with dualism that there is more to reality than described by fundamental physics. What this "more" amounts to depends on the precise type of emergentism one holds. We will get to that in due time. For now, take it as a promissory note that some version of emergence allows for agents with the capacity for free-will choices. This is the more challenging case to make since one has to show that some entities are composed of material stuff but are not reducible to material stuff. It will, in fact, require most of the rest of this book.

We should note that philosophers also discuss *property* dualism, in which there is one sort of physical substance with two different sorts of properties, some physical and some mental (Zimmerman 2010, 121–23). I will not pursue that idea further since it seems to me that most of what is interesting about property dualism has been subsumed under emergence.

Dualists may wonder why we are spending so much time on emergence when they have a solution to the problem of determinism. There are two reasons. First, a better understanding of reduction/emergence allows one to see exactly what's wrong with the skeptical claims in Chapter 1. Second, there are longstanding objections to dualism, including how immaterial minds interact with material bodies. The interaction problem and two others discussed in the next chapter are enough to make dualism a non-starter for many. Dualism might provide an out for the libertarian, but one with considerable baggage.

Two science-based arguments against dualism will be discussed in Chapter 5. Before turning there, we pause to consider how a different kind of determinism presents a challenge to libertarian choice.

Notes

1 There are more than just these two, but I take these to be the most prominent and promising.
2 The idea, once again, is that since no one has power over facts about the past or the laws of nature, and these factors determine all future events under determinism, no one has power over any facts of the future, including their own actions.
3 Why not start with the Big Bang itself? Because there is no instant in time or state of the universe *at* the Big Bang. In classical Big Bang cosmology, the equations wind back to a singularity, a point at which those equations no longer make physical sense. One needs to move away from the singularity just a bit for the notion of state to apply.
4 We will consider space-time and the universe itself in Chapter 4.
5 There are complications. Whether a system counts as deterministic or not sometimes depends on which notion of state one chooses (Butterfield 2005, sec. 4).
6 Momentum in physics has more components than one might guess. It includes the mass of the object, the direction it is moving, and its speed.
7 Phase spaces can be used to represent more than just systems of particles.
8 This way of putting it assumes that the system is conservative. There is no friction or other sort of dissipation in play.
9 "Order" here has to do with how many times a derivative is taken. Velocity is the derivative of distance, so the equation would be . Acceleration is the derivative of velocity, and so a second derivative of distance.
10 Kane makes a related point in Kane (1986, 121–22).
11 The Euler strut is a favorite of Mark Wilson, although Wilson himself believes these examples indicate "descriptive holes" rather than breakdowns in determinism (Wilson 2009, 181; Wilson 1989, 76).
12 This means that certain physical quantities would become infinite or undefined. Singularities can be produced even without collisions, as mathematician Zhihong Xia proved (1992). Also, see Earman (1986, 34).
13 van Strien mentions two strategies that were used to avoid this problem (2021, 7). First, since the odds of actual point particles colliding are small, some physicists thought we could just ignore this possibility. Second, Boscovich himself posited a short-range repulsive force between particles that would keep them from coming too close. Like Norton's Dome, these are strategies for getting around the indeterminism that presents itself in the mathematics.
14 Each of the three passages was translated by van Strien (2021, 14, 15, 17), respectively.
15 As Daniel Speak rightly points out (private correspondence), the Consequence Argument does not affirm determinism. It instead purports to show one of its implications.
16 Kane appeals to chaos theory as a possible amplifier of quantum events (2005, 134). Small changes in a chaotic system can produce dramatically different changes of state as a system evolves—the so-called "butterfly effect." This proposal has several problems, however, the foremost being that nature, including

the brain, is far less chaotic than the popular literature sometimes portrays (Koperski 2020, sec. 3.2).

17 van Strien argues that, contra the conventional wisdom, restoring determinism was never Bohm's goal (2025).

18 Hans Halvorson argues that violating quantum mechanical laws is not the trump card that Esfeld makes it out to be (2011, 160–62.) The GRW interpretation is also committed to a violation of standard quantum mechanics by positing a stochastic collapse of the wave function. Perhaps the action of immaterial minds, he speculates, likewise produces exceptions.

19 There are other non-physicalist options. See Mørch (2024) for more.

20 We will consider the options in Chapter 6.

21 Here is one such passage from Descartes's *Principles of Philosophy*, "All the particular causes of the changes which bodies undergo are covered by this third law—or at least the law covers all changes which are themselves corporeal. I am not here inquiring into the existence or nature of any power to move bodies which may be possessed by human minds, or the minds of angels…" (Descartes 2006, 1:242). As Daniel Garber sums up, "[These] passages suggest that in the *Principia* Descartes, at very least, left open the possibility that the activity of minds is not constrained by the laws of nature that hold for bodies" (Garber 1983, 116).

22 To cite one example: "If physicalism is true and a person is just a living human body, then the libertarian account of freedom would seem to be ruled out: events in physical systems can be rigidly caused or to some degree random, but a physical event can hardly be free in the libertarian sense. So libertarian freedom requires dualism of some sort" (Robinson 2011, 175).

Works Cited

Ahnert, Thomas. 2014. "Soul and Mind." In *The Routledge Companion to Eighteenth Century Philosophy*, edited by Aaron Garrett, 297–319. Routledge Philosophy Companions. London: Routledge.

Alembert, Jean le Rond d'. (1743) 2009. *Traité de Dynamique*. Whitefish: Kessinger Publishing. https://books.google.com/books?id=h0UhPwAACAAJ.

Arnold, Vladimir I. 1988. *Dynamical Systems III*. Berlin: Springer.

Becker, Adam. 2018. *What Is Real? The Unfinished Quest for the Meaning of Quantum Physics*. New York: Basic Books.

Bishop, Robert C. 2023. *Chaos Theory: A Quick Immersion*. New York: Tibidabo Publishing.

Bohm, David. 1952. "A Suggested Interpretation of the Quantum Theory in Terms of 'Hidden' Variables." *Physical Review* 85 (2): 166–79. https://doi.org/10.1103/PhysRev.85.166.

Boltzmann, Ludwig. 1905. *Populäre Schriften*. Leipzig: Verlag von Johann Ambrosius Barth.

Breidert, Wolfgang. 2008. "Leonhard Euler and Philosophy." In *Leonhard Euler: Life, Work, and Legacy*, edited by Robert E. Bradley and Charles Edward Sandifer, 97–108. Studies in the History and Philosophy of Mathematics, v. 5. Amsterdam: Elsevier.

Butterfield, Jeremy. 2005. "Determinism and Indeterminism." In *Routledge Encyclopedia of Philosophy*. 2nd ed. London: Routledge. https://doi.org/10.4324/9780415249126-Q025-2.

Campbell, Lewis, and William Garnett. 1882. *The Life of James Clerk Maxwell with a Selection from His Correspondence and Occasional Writings and a Sketch of His Contributions to Science*. London: Macmillan and Company.

Carroll, Sean. 2016. *The Big Picture: On the Origins of Life, Meaning, and the Universe Itself*. New York: Dutton.

Carroll, Sean. 2019. *Something Deeply Hidden: Quantum Worlds and the Emergence of Spacetime*. New York: Dutton.

Descartes, René. 2006. *The Philosophical Writings of Descartes*. Edited by John Cottingham, Robert Stoothoff, and Dugald Murdoch. Vol. 1. Cambridge: Cambridge University Press.

Earman, John. 1986. *A Primer on Determinism*. Dordrecht: D. Reidel Pub. Co.

Earman, John. 2004. "Determinism: What We Have Learned and What We Still Don't Know." In *Freedom and Determinism*, 21–46. Cambridge: MIT Press.

Esfeld, Michael. 2020. *Science and Human Freedom*. Cham: Palgrave Macmillan.

Fletcher, Samuel C. 2012. "What Counts as a Newtonian System? The View from Norton's Dome." *European Journal for Philosophy of Science* 2 (3): 275–97.

Fox, Robert. 1974. "The Rise and Fall of Laplacian Physics." *Historical Studies in the Physical Sciences* 4: 89–136. https://doi.org/10.2307/27757328.

Fraser, Craig. 1985. "D'Alembert's Principle: The Original Formulation and Application in Jean d'Alembert's Traité de Dynamique (1743)." *Centaurus* 28: 31–61.

Garber, Daniel. 1983. "Mind, Body and the Laws of Nature in Descartes and Leibniz." *Midwest Studies in Philosophy* 8: 105–33. https://doi.org/10.1111/j.1475-4975.1983.tb00462.x.

Goetz, Stewart. 2017. "Free Will and Substance Dualism." In *Routledge Companion to Free Will*, edited by Kevin Timpe, Meghan Griffith, and Neil Levy, 522–30. London: Routledge.

Goldstein, Sheldon. 2017. "Bohmian Mechanics." In *Stanford Encyclopedia of Philosophy*, edited by Edward N. Zalta. Metaphysics Research Lab, Stanford University. https://plato.stanford.edu/archives/sum2017/entries/qm-bohm/.

Halvorson, Hans. 2011. "The Measure of All Things: Quantum Mechanics and the Soul." In *The Soul Hypothesis: Investigations into the Existence of the Soul*, edited by Mark C. Baker and Stewart Goetz, 138–63. New York: Continuum.

Honderich, Ted. 1993. *How Free Are You? The Determinism Problem*. 1st ed. Oxford: Oxford University Press.

Hossenfelder, Sabine. 2020. "You Don't Have Free Will, but Don't Worry." *BackReAction* (blog). October 10, 2020. http://backreaction.blogspot.com/2020/10/you-dont-have-free-will-but-dont-worry.html.

Kane, Robert. 1986. "Principles of Reason." *Erkenntnis* 24 (2): 115–36.

Kane, Robert. 2005. *A Contemporary Introduction to Free Will*. Fundamentals of Philosophy Series. Oxford: Oxford University Press.

Koperski, Jeffrey. 2020. *Divine Action, Determinism, and the Laws of Nature*. New York: Routledge. https://doi.org/10.4324/9780429029110.

Lagrange, Joseph-Louis. (1811) 1997. *Analytical Mechanics*. Edited by Auguste Boissonnade and Victor N. Vagliente. Dordrecht: Springer Netherlands. http://link.springer.com/10.1007/978-94-015-8903-1.

Laplace, Pierre Simon. (1814) 1902. *A Philosophical Essay on Probabilities*. Translated by F. W. Truscott and F. L. Emory. Chichester: Wiley.

Leibniz, Gottfried Wilhelm. (1714) 1898. The Monadology and Other Philosophical Writings. Translated by Robert Latta. Oxford: Clarendon Press.

Leibniz, Gottfried Wilhelm. (1765) 1916. *New Essays Concerning Human Understanding*. Translated by A. G. Langley. Chicago: Open Court Publishing Company.

Mach, Ernst. 1905. *Erkenntnis Und Irrtum: Skizzen Zur Psychologie Der Forschung*. Leipzig: Verlag von Johann Ambrosius Barth.

Marion, Jerry B., and Stephen T. Thornton. 1988. *Classical Dynamics of Particles & Systems*. 3rd ed. San Diego: Harcourt Brace Jovanovich.

Mørch, Hedda Hassel. 2024. *Non-Physicalist Theories of Consciousness*. Cambridge: Cambridge University Press. https://doi.org/10.1017/9781009317344.

Norton, John. 2008. "The Dome: An Unexpectedly Simple Failure of Determinism." *Philosophy of Science* 75 (5): 786–98.

Pereboom, Derk. 2022. *Free Will*. Elements in Philosophy of Mind. Cambridge: Cambridge University Press. https://doi.org/10.1017/9781108982511.

Pitts, J. Brian. 2019. "The Mind-Body Problem and Conservation Laws: The Growth of Physical Understanding?" http://philsci-archive.pitt.edu/16380/.

Poincaré, Henri. (1913) 1917. *Dernières Pensées*. Paris: Flammarion.

Robinson, Howard. 2011. "Substance Dualism and Its Rationale." In *Free Will and Modern Science*, edited by Richard Swinburne, 158–77. Oxford: Oxford University Press.

Sapolsky, Robert M. 2023. *Determined: A Science of Life without Free Will*. New York: Penguin Press.

van Strien, Marij. 2014a. "On the Origins and Foundations of Laplacian Determinism." *Studies in History and Philosophy of Science Part A* 45: 24–31.

van Strien, Marij. 2014b. "The Norton Dome and the Nineteenth Century Foundations of Determinism." *Journal for General Philosophy of Science* 45 (1): 167–85. https://doi.org/10.1007/s10838-014-9241-0.

van Strien, Marij. 2021. "Was Physics Ever Deterministic? The Historical Basis of Determinism and the Image of Classical Physics." *The European Physical Journal H* 46 (8): 1–20. https://doi.org/10.1140/epjh/s13129-021-00012-x.

van Strien, Marij. 2025. "Why Bohm Was Never a Determinist." In *Guiding Waves In Quantum Mechanics: 100 Years of de Broglie-Bohm Pilot-Wave Theory*, edited by Andrea Oldofredi. Vol. 45. Oxford: Oxford University Press.

Wallace, David. 2012. *The Emergent Multiverse: Quantum Theory According to the Everett Interpretation*. Oxford: Oxford University Press.

Wilson, Mark. 1989. "Critical Notice: John Earman's a Primer on Determinism." *Philosophy of Science* 56 (3): 502–32.

Wilson, Mark. 2009. "Determinism and the Mystery of the Missing Physics." *British Journal for the Philosophy of Science* 60 (1): 173–93.

Wilson, Mark. 2016. "Mechanics, Classical." In *Routledge Encyclopedia of Philosophy*, 1st ed. London: Routledge. https://doi.org/10.4324/9780415249126-Q068-1.

Xia, Zhihong. 1992. "The Existence of Noncollision Singularities in Newtonian Systems." *The Annals of Mathematics* 135 (3): 411. https://doi.org/10.2307/2946572.

Zimmerman, Dean. 2010. "From Property Dualism to Substance Dualism." *Proceedings of the Aristotelian Society, Supplementary Volumes* 84 : 119–50.

4 Relativity and the Block Universe

4.1 A Different Sort of Determinism

There is a physics-based challenge to free will that was not raised by the free-will skeptics in Chapter 1. Instead of determinism and the laws of physics, the issue in this chapter is the nature of the future itself. Does the future literally exist? Is it fixed, like the past, or yet to be determined? Can a decision made today influence how that future will unfold?

Leeway libertarianism (Section 2.3.1) demands that agents possess genuine alternative possibilities, which seems to require an open, undetermined future. Yet special and general relativity appear to present a direct challenge to this requirement, suggesting instead that the future exists as a fixed, unchanging reality within the structure of spacetime. This chapter investigates whether relativity genuinely denies the dynamic, temporally open future that libertarian free will appears to demand. Through analysis of competing interpretations of relativistic physics, we'll determine whether the theory necessarily eliminates the temporal contingency essential to libertarian accounts or whether alternative readings preserve the conceptual space required for genuine agency.

The challenge of relativity will be clearer if we first consider some issues in the philosophy of time. There are two main camps: presentism and eternalism. The former takes the flow of time much like we experience it. The latter takes that experience to be something of an illusion. We will first consider how eternalism poses a problem to libertarian freedom. After that, we will see whether relativity entails eternalism and refutes presentism—a question that will take up most of the chapter.[1]

4.2 The Philosophy of Time

While there are several well-known positions on the nature of time (Miller 2016), only two will be taken up here. The primary difference between

DOI: 10.4324/9781003681199-4

them is about ontology. Does the future exist? Presentism says "no"; eternalism says "yes."

According to presentism, references to "the past" or "the future" are mere figures of speech (Ingram and Tallant 2023; Mozersky 2011; Zimmerman 2011). There are no objects or events "in the past" or "in the future" because only the present exists. Time travel is impossible not because we lack technology but because there is literally nowhere to go. Presentists also believe there is a flow of time. The present or the "now" continually moves along, much as it seems.[2]

Eternalism, which takes what we think of as past and future to be fully real (Mellor 1998; Sider 2001), goes back at least as far as Anselm (Rogers 2015, 111). Temporal reality is something like a movie reel. At any given place on the reel, we could talk about earlier scenes or later ones, but they are all part of the completed movie. While playing the movie on a projector gives the illusion of a moving present, no part of the reel has any more reality than any other. Similarly, eternalism holds that the whole of time fully exists. From a God's-eye point of view—seeing reality as it is—there is no changing present.[3] Any feeling we might have about the passage of time is purely a matter of subjective experience that does not correspond to objective reality. Even one's experience is frozen in place.

If you're interested in the nature of reality (metaphysics), this is a fascinating dispute. Can it really be that there is no passage of time? If so, how can we make sense of our feeling or experience of an ever-changing "now"? And if only the present exists, then in what sense are claims about the past or future true (Miller 2016, 354–56)? True statements are thought to have a truth-bearer (a sentence or proposition) and a truth-maker (something in reality that makes the sentence or proposition true). For example, the statement "Mars is the fourth planet in our solar system" is true, and it has a truthmaker: the planet that we named "Mars" and its orbit relative to the others. But what about "Vikings visited the Hagia Sophia in the ninth century?" That also appears to be true, but under presentism, nothing in the ninth century currently exists. Hence, the true claim lacks a truthmaker. Is that a problem? Again, if you like a metaphysical dispute, this is a good one. While intuitions typically side with presentism, a plurality of philosophers leans the other way (Bourget and Chalmers 2023, 8).

For our purposes, the questions just raised are rabbit trails that we can safely ignore as we focus on the implications of presentism and eternalism for free will. Suppose we call this present moment t_1. Now consider some choice to be made at a later date, like what you will have for dinner in exactly one month. Call that time t_2. According to eternalism, the events at t_2 are just as real as the events at t_1. Every event up and down the timeline has the same ontological status. The "future"—which again is just a figure of speech—does not have some lesser, shadowy existence than the present.

Just like each frame in a movie, it's all there—beginning, middle, and end. Moreover, once the movie has been produced, there is no contingency. I am not sure how many times my kids watched the original *Star Wars*, but there was no sense in which Luke Skywalker might not blow up the Death Star in the end. Likewise for time under eternalism. Say you have pizza at t_2. Is there anything you can do now to change that? No. You having the pizza just *is*, like an insect in amber. The timeline—past, present, and future—likewise unchangeably exists.

That seems to present a problem for libertarian freedom. You do not now have the ability to change how the future unfolds. "The future" is just as fixed as the past according to eternalism. It might feel like there is contingency about how future events will play out, but all the events at t_2, like all times, fully exist and cannot be changed.

Thus far, it's an interesting philosophical question. What does physics have to say about it?

4.3 Special Relativity and Eternalism

One theory in particular challenges our intuitions about time: special relativity (SR). A key claim of SR is that no matter how fast one is moving towards or away from a source of light, the observed speed of that light remains the same. Yet other things don't behave that way. For example, two cars on the highway in the same lane traveling at the same speed appear not to be moving with respect to one another. If they were close enough, a person in one car could lean out the window and hand something to a person in the other car. It would be the same as if both cars were sitting in a driveway. In contrast, cars moving in the opposite lane seem to be moving very fast. Two cars moving at 65 kilometers per hour in opposite directions appear to be moving at 130 km/hr from the point of view of each other. This is the sort of vector addition one learns in freshman physics.

Light, however, behaves differently. According to SR, light has the same speed c (3×10^8 meters per second)[4] whether the source of that light is moving toward an observer or away. Two spaceships approaching one another at $0.8c$ would each observe the light emitted from each ship to be moving at c. The vector addition of velocities does not apply. While that might sound like a small exception, it has surprising consequences for other sorts of measurements. For example, consider a group of objects moving in the same direction at the same speed. That group constitutes an *inertial reference frame*. Me and everything in my office is one such frame.[5] A group of astronauts halfway to Mars is another. To introduce more jargon, since c is the same for everyone, we say it is *invariant* across reference frames. Something that is not invariant according to SR is length.

From the point of view of one inertial frame, everything in another seems slightly compressed along the direction it is traveling.[6] This effect is known as *length contraction*. Why does it happen?

Mathematician Hermann Minkowski provided the now-standard way of understanding this and other oddities of SR. Students learn to specify positions in space using Cartesian coordinates (x, y, z). Minkowski combined this three-dimensional space with a linear axis of time to create four-dimensional spacetime (x, y, z, t). Each point in spacetime is called an *event*. Consider a person in an inertial frame within a spacetime coordinate system (Figure 4.1). The person is the point labeled O. The time coordinate is on the vertical axis. While there should be three spatial coordinates, one has been left out for simplicity. Events at coordinates other than O fit into one of three regions. The edges of the so-called *light cone* are defined by particles moving at the speed of light. Particles arriving at O but moving slower than c are in the past light cone. Those leaving O are in the future light cone. Events within those cones are called *timelike*.

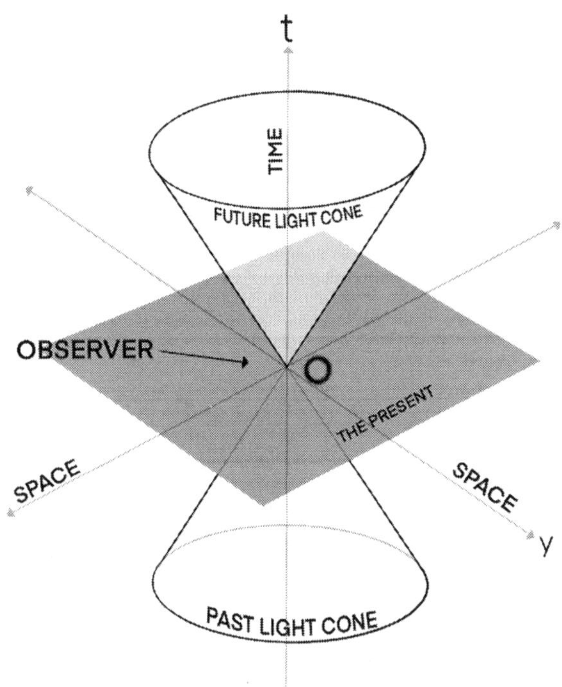

Figure 4.1 Light cone.

Source: Artwork by Ashley McNutt-Greene. Used with permission.

Those outside the cones are called *spacelike*, which are too far from O to be reached even by light.

An important idea here is that the light cone structure is invariant: observers in other inertial frames agree on the speed of light and the regions of each other's light cones. More technically, they agree on the spacetime *interval*: $s^2 = (x_2 - x_1)^2 + (y_2 - y_1)^2 + (z_2 - z_1)^2 - c^2(t_2 - t_1)^2$, which will remind some readers of the Pythagorean Theorem. An interval in spacetime is analogous to distance in space, but here the separation is between events in spacetime rather than places in space. The interval is invariant, so observers in every inertial frame will agree on its value.[7] But if they agree on this, they will disagree about other things. For example, they will disagree about the passage of time in different inertial frames (*time-dilation*). This is often described as clocks in other inertial frames moving too slowly. In fact, it's the passage of time itself that seems to run too slowly: clocks, heartbeats, radioactive decay ... literally everything going on in the other inertial frame. Why? Because the elapsed time between two events, $t_2 - t_1$, is only part of the interval, and the parts are not invariant. Observers will disagree about how much time has passed.

It gets worse. Consider another event E that, according to the person at the origin, is simultaneous with O. From the point of view of observers in some other inertial frames, E happens before O. According to yet other inertial frames, O happens before E. This "relativity of simultaneity" means events that are in the future according to one observer are in the present according to another.

That takes us back to the philosophy of time. Presentists about time claim that only events in the present exist. The past and future, they say, are not real. If we combine that with what we just said about observers in different inertial frames, they disagree about which events are real. This is not about some people simply having a better sense of time than others. If the distances are on the order of galaxies, the two observers will disagree about whether *several days* worth of events have already occurred (Penrose 1989, 201). According to the presentist, they disagree about what exists.

Of course, people disagree about all sorts of things. If two people disagree about the atomic number of iron, at least one of them is wrong. Disagreements in SR, however, are not due to a lack of knowledge. No inertial frame is special, and no observer has any better claim to having the right answer about time. Nobody has *the* right answer because there is no single objective right answer about the passage of time. Recall that the elapsed time $t_2 - t_1$ is not invariant. The duration between events varies from the point of view of different inertial frames. In contrast, there *is* a right answer about the interval. The interval is invariant, incorporating both space and time, not just time itself.

Why think there is no right answer about the passage of time? Perhaps the truth is just inaccessible, like precisely how many stars are in a nearby galaxy. The answer goes back to invariance. Properties that differ from one perspective to another are often deemed less "real" than those that do not. For example, the sweetness of wine can vary from person to person. However, the volume of wine in a glass remains the same regardless of who measures it. This objective, invariant property is deemed more real.

In SR, observations related to spatial distances and the passage of time vary depending on the observer's inertial frame. The spacetime interval between two events, in contrast, is invariant. No matter the observer's frame of reference, the interval remains constant. Physicists consider these invariant quantities, like the interval, to be more fundamental or "real" because they do not change with perspective. This leads to the conclusion that four-dimensional mathematical constructs, which include both space and time dimensions, are the true descriptors of reality.

A key point for the presentist-eternalist debate is that observers in different inertial frames disagree about which events are in the past, present, and future. As we have seen, events considered to be in the future from one frame of reference may be happening "now" in another. These temporal judgments are not invariant (even though the interval as a whole is). Since every inertial frame is ontologically on a par with every other, if a presentist in one frame accepts all events that are simultaneous with their observed "now" as real, there is no principled reason to reject those in other reference frames from doing the same. Without a privileged inertial frame—one that has the right answer—all temporal judgments must be given equal weight. "Each observer has his own set of 'nows,' and none of these various layers can claim the prerogative of representing the objective lapse of time" (Gödel 1949, 558). If some events are in your present, then I must consider those events to exist, even if those events have not yet happened from my point of view. This means that events in my future exist, which is precisely what presentism denies.

So which future events should we count as real? The eternalist has an answer: all of them. The whole of spacetime exists. Spacetime facts are what the (invariant!) interval refers to. As mathematician Roger Penrose put it, "It begins to seem that if anything is definite at all, then the entire spacetime must indeed be definite! There can be no 'uncertain' future. The *whole* of spacetime must be fixed, without any scope for uncertainty" (1989, 304). Sean Carroll agrees:

> The past and future are equally real. This isn't completely accepted, but it should be. Intuitively we think that the "now" is real, while the past is fixed and in the books, and the future hasn't yet occurred. But physics teaches us something remarkable: every event in the past and future is implicit in the current moment. This is hard to see in our everyday lives, since we're nowhere close to knowing everything about the universe

> at any moment, nor will we ever be — but the equations don't lie. As Einstein put it, "It appears therefore more natural to think of physical reality as a four dimensional existence, instead of, as hitherto, the evolution of a three dimensional existence."
>
> (Carroll 2011)

Irreconcilable disagreements between inertial frames reveal that time is merely a matter of our perspective. In fact, on this view, there is no passage of time:

> Such a view can be formalized in the idea of a *block universe*: space and time are represented as merged into an unchanging spacetime entity, with no particular space sections identified as the present and no evolution of spacetime taking place. The universe just *is*; a fixed spacetime block, representing all events that have happened and that will happen. This representation implicitly embodies the idea that time is an illusion; time does not *roll on* in this picture. All past and future times are equally present, and the present *now* is just one of an infinite number.
>
> (Ellis and Goswami 2014, 244)[8]

The idea of a "block universe" mentioned here is common in the philosophical literature. From a God's-eye perspective, "The objective world simply *is*, it does not *happen*" (Weyl 1963, 116). This scientific verdict strongly favors eternalism.[9]

What might a presentist say at this point? Perhaps the most common interpretation of SR is wrong. Perhaps there is, in fact, a preferred inertial frame, although there is no way to physically detect it. This would not be unheard of in physics. There are other undetectable facts, such as the internal dynamics of black holes, the precise nature of the Big Bang, and the geometry of space beyond the observable universe. Perhaps the moving present is another.

The problem with saying that the moving present is undetectable is that a big motivation for presentism is that it fits with how things seem, i.e., that we do sense it to some degree (Miller 2016, 353; Eagle 2021, 94). It feels like there is a passage of time and something special about the present. But if that special moment is undetectable, how can it be responsible for how things seem to us? If presentists claim there is a fact of the matter about "now" but grant that it is impossible to sense its passage, that would undermine the motivation for presentism itself. This means that the presentist will have to do more than allude to undetectable facts.

While the presentism-eternalism debate is interesting in itself, the central concern of this book is free will. While fewer philosophers discern a clear connection between free will and SR, some make it explicit:

> [There] does not exist an event, that is not already in the past for some possible distant observer, at the (our) moment that the latter is "now" for us. Such event is as legitimately past for that distant observer as is the moment five minutes ago on the sun for us ... [We] may conclude that there is determinism ... Also there is no free will ...
>
> (Rietdijk 1966, 343)

> In the Minkowski four-dimensional world ... there is no free will, since the entire history of every object is realized and given once and for all as the object's worldtube [roughly its path through spacetime].
>
> (Petkov 2009, 172)

> Ultimate moral responsibility is also ruled out by the theory of relativity. Einstein himself, in a piece written as a homage to the Indian mystical poet Rabindranath Tagore, said that "a Being endowed with higher insight and more perfect intelligence, watching man and his doings, would smile about man's illusion that he was acting according to his own free will."
>
> (Sommers 2003)

The future is fixed in the block universe. You do not have it in your power today to make a choice that will influence how later events will unfold. What we think of as "the future" fully exists, and those events cannot be changed. It seems that libertarianism tacitly assumes that presentism (or something in the neighborhood)[10] is true. Most libertarians believe that choices made at one point in time can change which events take place later. In other words, the future is open rather than fixed.

Eternalism therefore seems to be a problem for libertarian freedom. If what we think of as the future exists and does not change, we are not now free to opt for a different course. Is this determinism? Of a sort (Rietdijk 1966). Nothing in this picture says that one's choices are somehow brought about by the laws of nature or some previous set of causes. So eternalism is not a kind of physical determinism, which is what most libertarians are worried about. In fact, it looks more like theological determinism, in which God's knowledge of the future is absolute (Section 2.2.1.1). Say that at time t_1, God knows that an agent at t_2 will buy a book. The agent cannot alter the events at t_2; otherwise, God would have had a false belief at t_1. Nonetheless, God is not making the agent buy the book. Eternalism presents similar issues, but instead of God's foreknowledge, the problem is grounded in actually existing facts about the future that cannot be changed. In neither case can the agent do otherwise.

This inability presents a problem for libertarians, or at least some of them. Yet some libertarians are not bothered by eternalism and the block universe.

4.4 Sourcehood to the Rescue?

Recall a distinction made in Section 2.3 between leeway and sourcehood. Leeway libertarians endorse AP: A "person has free will only if he could have done otherwise" (Timpe 2012, 70), i.e., if he has the ability to do other than what he actually does. AP implies or assumes an open future with forking paths. Sourcehood libertarians instead emphasize control over AP. The issue is not about one's ability to do otherwise but whether my free will "places my stamp upon the world, and whether it is up to me or not to have that particular stamp or some other as my mark upon the world" (McKenna 1997, 73–75). My actions, my "stamp," are events within my control.

Eternalism conflicts with leeway libertarianism because there are no forking paths in a block universe. The future—in fact, the whole of time—contains exactly one fixed path. Sourcehood, on the other hand, does not require an open future (Page 2022, 86). Source theorists need not hold any particular position about the nature of time. The only question is whether an agent's free choice is a necessary condition for which events occur at a point in time. That there is a fact of the matter about whether an agent has pizza next Friday night is not a problem. Free will, says the source theorist, only requires that *one's choice* in the moment determines whether the agent orders the pizza. If it were determined instead by something external to the agent, like the laws of nature or manipulation by a mad scientist, then that person would not be free. The exercise of undetermined choice is the issue, not the structure of time.[11] In short, eternalism does not conflict with sourcehood libertarianism.[12]

As we saw in Chapter 2, many libertarians do not accept sourcehood. More precisely, they think that it is not sufficient for libertarian freedom. For them, the fixedness of the eternalist future is a problem, and so then is SR. Leeway libertarians need presentism or some other view of time that allows for an open future.[13] In the rest of this chapter, we will consider what resources presentism has for rejecting the block universe.[14] We first consider some ways in which philosophy of science might help.

4.5 Idealizations and Antirealism

A longstanding debate among philosophers of science is the question of *scientific realism* (Okasha 2016, chap. 4). While there are different ways to construe it, realists tend to take mature theories in science to be true (or at least approximately so), where truth is something like "corresponds to the way physical reality is." To say that electrons have a negative charge is to assert the existence of electrons, even though they are too small to be observed, and that they have the property of being negatively charged. Realists take mature science to be telling us what physical reality is like. Most scientists and philosophers are scientific realists of some variety or other.

Antirealists also have a high view of science but deny that it is in the business of providing true descriptions of reality. *Instrumentalists*, for example, take the aim of science to be making successful predictions and technological advances.[15] Theories don't need to be true in order to do that. Ptolemaic astronomy worked well—better, in fact, than Copernicus's earliest models—but it was not true. Many twentieth-century physicists were trained to think in instrumentalist terms. They were told not to consider what the equations "really mean." All that matters is that they work. More philosophically inclined graduate students asking questions about, say, the interpretation of quantum mechanics were famously told to "shut up and calculate."

Presentism could be construed as a kind of antirealism about relativity: SR works but does not provide a true picture of the nature of time. That, however, seems rather extreme. Relativity is considered one of the greatest scientific achievements of the last century. Presentists trying to preserve their position by retreating to antirealism look ad hoc.

On the other hand, no one is a realist about everything that appears in textbook science. For one thing, physicists and engineers employ a variety of idealizations in their work. Students are told they can consider sliding blocks to be frictionless, common electric circuits to be resistance-free, and gas containers to have an infinite number of particles. The models used in general relativity (GR) often treat stars and even galaxies as mathematical points. Everyone knows these idealizations are not literally true. Moreover, many theories, laws, and models have limited domains of applicability. Most engineers learn very little about quantum mechanics and relativity, as classical mechanics accurately describes the physics encountered in our everyday environment. Classical mechanics only breaks down at very small scales and when objects are moving close to the speed of light.

How might presentism make use of this? Instead of taking the whole of SR in an antirealist way, perhaps some aspects of it could be. One might be an antirealist about the simultaneity of distant events, for example. We do not have empirical access to events that are spacelike separated, outside one's light cone. To say that those events are simultaneous with O (Figure 4.1) is something one calculates to be the case, not something we experience, as John Lucas explains:

> The lines of simultaneity for a given inertial frame of reference do not determine what is currently going on at distant places, but only what dates should be ascribed to them in order to make electromagnetic phenomena coherent. The Special Theory is a theory of electromagnetic radiation and determines distances and durations, and hence positions and dates, by means of light signals. As far as electromagnetic phenomena are concerned, we have no other means of telling exactly when a distant event ... takes place... The ascription of presentness, pastness, or futurity, to events outside the light cone is nominal rather than real, and has no bearing on their ontological status.
>
> (Lucas 2008, 284–85)

Perhaps there is no fact of the matter regarding the simultaneity of distant, spacelike-separated events. (See Figure 4.1 again for spacelike and timelike regions relative to an event.) Ellis and Goswami suggest something like this:

> The concept of simultaneity is only physically meaningful for neighboring events; it has no physical impact for distant events, it is merely a theoretical construct we like to make in our minds. What we think is instantaneous makes no difference to our interaction with a vehicle on Mars.
>
> (Ellis and Goswami 2014, 250)

They take the simultaneity of distant events to be a distraction (2014, 261). What matters are interactions between objects, which happen within timelike regions of light cones (or their edges when it comes to light). While one can calculate which distant, spacelike-separated events should count as simultaneous with some event O, such calculations are not physically meaningful. The upshot is that many of the examples offered earlier giving conflicting views about whether distant events have already occurred are not describing physical reality. One ought not to infer anything about the nature of time from such examples.

Presentists could also grant that in terms of inertial frames and relative velocities, SR is correct, and a block universe is its most straightforward interpretation. But there are many aspects of science that SR ignores, such as thermodynamics and biological systems, in which the passage of time plays a key role (Bishop and Atmanspacher 2011, 106). One might say that SR applies to an idealized, limited domain and ought not be considered the whole story regarding science and time. Its models

> do not apply to spacetimes including complex systems because the equations of state they assume are too simple—they do not include

> friction and dissipative effects, hierarchical structures, feedback effects, or the causal efficacy of information, and they do not take quantum uncertainty into account.
>
> (Ellis 2007, 74)

The quantum case is particularly interesting. Some physically distant (spacelike separated) events do happen simultaneously, according to quantum mechanics. When particles are *entangled*—a technical term—interactions with one have an instantaneous effect on the other no matter how far apart they are (Maudlin 2002, 23–24). There would therefore be a fact of the matter about the simultaneity of such events.[16]

According to Dean Zimmerman (2011, 220), none of this amounts to a rejection of SR, as some eternalists say (Sider 2001, 52). Presentists should accept SR the way they accept Kirchhoff's circuit laws. Neither is false, but neither tells a complete story. Einstein made a similar move in his rejection of action-at-a-distance (Maudlin 2008, 156–57). Whereas GR is silent on local vs. non-local interactions, Einstein believed that nature only allows the former: Distant objects cannot have an immediate causal influence on one another. In other words, there are physical truths that GR does not account for. Presentists are doing much the same with respect to the passage of time. They believe reality has more temporal structure than one finds in the block universe, and so there are temporal facts SR fails to capture. In the jargon, presentists believe there is a *preferred foliation* of spacetime—a fact of the matter regarding the passage of time.

Eternalists reply that this additional temporal structure is unnecessary (Eagle 2021, 95). There is no evidence for a preferred foliation. Examples like thermodynamics and biological systems that are supposed to imply temporal passage can be dealt with in eternalist terms. The only known sense in which SR is idealized is that it ignores gravity, hence the need for GR. But both SR and GR, says the eternalist, entail a block universe.

To see whether this last claim is true requires a deeper dive into the philosophy of physics. While the technical details will be kept to a minimum, there's no way to avoid them all at this point.

4.6 Two Approaches to Special Relativity

One dispute among philosophers of physics is whether spacetime or the relevant laws of nature are more fundamental (Brown 2005; Brown and Read 2021; Read 2023, 62–63). The standard, geometrical approach to SR says that the structure of spacetime explains the behavior of matter and the nature of the dynamical laws governing matter—SR is fundamentally about spacetime. The rival, dynamical approach says instead that spacetime arises from and is explained by the properties of the physical laws involved. Natural laws are more fundamental than spacetime. Einstein can be found on either side of this divide at different points in his life. He initially referred to the Minkowski spacetime approach as "superfluous learnedness," although he later warmed to the idea (Pais 2005, 152).

Before getting into the details, how might this alternative understanding of SR help the presentist? Although views differ, many take the dynamical approach to show that spacetime itself does not exist. There is no block universe, they say; there are only physical objects that behave in accordance with the laws of nature. While spacetime is an undeniably useful way of thinking about SR, it is not real or, at the very least, not a fundamental entity. But if there is no block universe, SR's support for eternalism likewise disappears.

Both the dynamical and geometrical approaches hold that the laws of physics are *Lorentz invariant*: The laws act the same within every inertial reference frame. (Recall that an inertial frame is simply a set of objects moving at the same, constant velocity.) They could be "at rest" in a lab, moving on a (non-accelerating) supersonic jet, or traveling at close to the speed of light. Lorentz invariance means that all the experiments and observations in one inertial frame would be indistinguishable from those in all other inertial frames. Lorentz invariance also guarantees that the interval is the same for all observers, which in turn leads to the distinctive SR phenomena such as length contraction and time dilation discussed earlier. The question is what ultimately grounds Lorentz invariance? Why are the laws that way?

The geometrical approach says it is the four-dimensional structure of spacetime (technically the metric) that determines Lorentz invariance. Observers in spacetime measuring duration or distance from different inertial frames only see part of this structure. The interval, in contrast, correctly describes the spatio-temporal distances between events. What we think of as the length of an object is a projection of an interval into different inertial frames, something like the familiar two-dimensional sketch of a three-dimensional cube (Figure 4.2). The sketch is a projection of a three-dimensional object onto a two-dimensional surface. The projection changes depending on the angle one looks at the cube, and so many different two-dimensional projections correspond to the same three-dimensional object. Likewise, under the geometrical approach, it is the actual, four-dimensional reality of spacetime that determines truths about the interval and the laws of physics (Balashov and Janssen 2003, 340).

> [The] laws are Lorentz invariant *because* spacetime has a Minkowski geometry, and the symmetries of the laws must match those of the spacetime geometry. The geometry of spacetime explains length

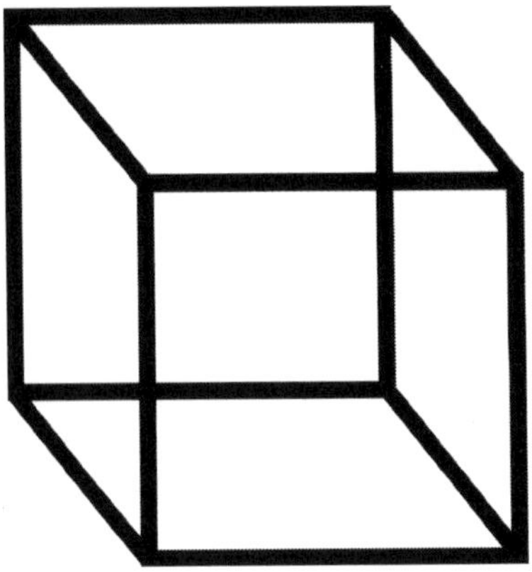

Figure 4.2 Cube projection.

> contraction because it places constraints on the structure of the dynamical laws.
>
> (Skow 2006, emphasis mine)

The dynamical approach says that the priority of explanation is the other way around. The laws of nature explain our observations. Spacetime—if there is such a thing—and its properties depend on the laws. This was the view of earlier pioneers of relativity, such as FitzGerald, Lorentz, Larmor, and Poincaré.[17] Einstein himself summed up SR in 1940 without referring to spacetime: "The content of the restricted relativity theory can … be summarised in one sentence: all natural laws must be so conditioned that they are covariant with respect to Lorentz transformations" (Brown and Read 2021, 70). (The last clause in this quote is another way of referring to Lorentz invariance.) When the speed of light is the same for all observers and the laws are Lorentz invariant, phenomena like time dilation, etc., are the result. The key is that such invariance is a property of the laws themselves. Nothing more fundamental makes the laws have this property (Brown and Read 2021, 77). There is nothing, say, about the composition of moving rods—plastic or iron—that would in turn require invariance in the laws. More importantly, there is nothing about geometry or spacetime that grounds Lorentz invariance. Spacetime and its structure explain nothing, nor do they fully dictate the behavior of material objects. One needs dynamical laws for that (Huggett, Hoefer, and Read 2023, sec. 4).

There is some debate about how all this maps onto an older controversy between substantivalism and relationalism about space itself (Read 2020, 25–27). That debate, which traces back to Newton and Leibniz, is about whether space has its own ontology over and above objects in spatial relations to one another (Hoefer, Huggett, and Read 2023). Newton believed that space is real. It is a substance all its own. This means that in principle there could be such a thing as empty space. Leibniz disagreed. He believed that there is no such thing as space itself. There are only objects that have different spatial relations to one another. The geometrical approach fits more naturally with substantivalism. The dynamical approach looks more like relationalism, although not all of its proponents agree (Huggett, Hoefer, and Read 2023, sec. 4).

Whether SR supports eternalism depends on which ontology is correct. If the geometrical approach is right and spacetime exists, that leads to the block universe. But if spacetime is merely a representation of what the laws dictate, then no reification is required. Both are live options with respect to the physics of SR. At the very least, this means that SR does not entail a block universe.[18]

To this point, we have only considered SR. But SR includes a significant idealization in that it ignores the presence of gravity. As such, we know that it does not accurately describe our universe. To do so, the discussion needs to move on to GR. The presentist has other options there. Note to the reader: things are about to get hairier. If you wish to skip the next two sections, the upshot is that GR provides the presentist some helpful options that are not available in SR.

4.7 General Relativity and Foliations

There are several varieties of time in relativity. The "time" in "spacetime" refers to *coordinate time*. This is represented by the t in the equation for the interval. The coordinate time between events differs from one inertial frame to the next. Another is *proper time*, which is measured along timelike curves. This is the local sense of time had by each observer or object. Some philosophers argue that proper time represents a mind-independent temporal becoming, contrary to the block universe (Savitt 2011, 563). Most, however, think that presentism needs a measure of time that applies to the universe as a whole, not just local observers. Global notions of time are available in GR, all of which rely on *foliations*—a technical term for a simple idea. It just means that one has taken a spacetime and then sliced it into temporal moments signifying a passage of time. These are better candidates for the sort of time presentists are looking for.

4.7.1 Cosmic Time

Einstein's field equations for GR have a kind of *symmetry*, a term used to describe a kind of sameness or lack of distinctive structure. A cue ball in billiards, for example, has rotational symmetry. It looks the same no matter how it is rotated. The structure lacking in Einstein's field equations is anything representing an overall ("global") notion of time. This is why GR seems to support eternalism. The underlying laws have nothing that describes the passage of time.

As Ellis argues, however, this inference bypasses a critical step. "This symmetry ... cannot be used to justify a Block Universe as a representation of the universe in which we live" (Ellis 2024, 4). Why not? Because the actual universe is represented by solutions (models) to the field equations. It is these models that describe real-world systems. The most commonly used models have more structure (and less symmetry) than the equations themselves (Ellis and Matravers 1995). Ellis has in mind the FLRW (Friedman-Lemaître-Robertson-Walker) models, which can be foliated in a natural way (Smeenk 2016, 205). The models treat matter in space as something like a fluid or dust. A set of "fundamental observers" can be defined who are at rest with respect to this matter and who are able to synchronize their clocks with one another in terms of local measurements.[19] The collection of fundamental observers and their synchronized clocks define *cosmic time*.

> Cosmic time in the FLRW models resembles Newtonian absolute time: for any two events in spacetime, the cosmic time elapsed between the events has a definite, observer-independent value. All of these properties lead to the conclusion that spacetime in the FLRW models breaks down into space and time much as in Newtonian spacetime, as Jeans and Eddington emphasized soon after the models were discovered.
>
> (Smeenk 2016, 205)

This sounds like excellent news for presentism. Unlike SR, the most common models used in GR yield an objective passage of time.

Nonetheless, the eternalist can point to some limitations. First, fundamental observers and cosmic time are just one foliation scheme among others yielding different measures of time (Smeenk 2016, 205). Second, not all FLRW models allow such foliations. Cosmic time requires space to be homogeneous and isotropic, meaning it looks the same in every direction. Observations of the night sky show that this is obviously not the case in our universe. We note that one set of constellations appears in one part of the sky and another set elsewhere. Moreover, if cosmic time depends on the arrangement of matter in space, then it is a contingent property. More realistic models, with galaxies and the like, resist foliation. Cosmic time, therefore, does not look like a fundamental aspect of nature, which the presentist claims it is.[20]

Presentists can answer this objection by noting that some foliation schemes are far more natural than others. By analogy, a fused deck of cards could be divided by separating each card from the others or by slicing diagonally across the entire deck. Only the first method preserves the properties of the cards. Cosmic time is similar. It carves spacetime in a more natural way than do the alternatives (Torretti 1996, 230). As for the idealizations involved, they are not unrealistic at cosmological scales, which is why cosmic time is commonly used. I have yet to hear any objections to cosmic time when it is used to determine the age of the universe, for example.

What about the contingent nature of cosmic time? Don't presentists believe that time applies to all physically possible worlds? Not necessarily. While it would be nice for the presentist if time turned out to be a part of every fundamental theory, it need not be. Our sense of the direction (or "arrow") of time likewise seems to be dependent, in this case, on the much lower entropy of the past. Perhaps presentists should be willing to give up the intuition that time is in some sense necessary, or at least time as we experience it in the actual universe.

In the end, cosmic time appears to be a live option for the presentist. It marks out an objective flow of time motivated by common solutions to Einstein's field equations. In this sense, GR is more favorable to presentism than SR.

Let's turn to another common foliation scheme physicists use to apply GR to real-world systems.

4.7.2 ADM 3+1

In this section, we will discuss two different mathematical versions of GR. Readers might wonder how there can be different versions of a single theory. It's more common than one might imagine. Schrödinger and Heisenberg developed two different sets of equations for quantum mechanics that were later shown to be equivalent. In classical mechanics, one can use Newton's laws and forces or the Lagrangian approach based on energy. Physicists use whichever is more convenient.

The mathematical core of GR is Einstein's field equations. There is nothing in those equations that represents the passage of time. This lack of temporal structure is what lends support to a block universe interpretation and eternalism. The field equations are not, however, the version of GR physicists use on a regular basis. For making calculations, they turn to the so-called "Hamiltonian" or "ADM 3+1" approach named for Richard Arnowitt, Stanley Deser, and Charles W. Misner (Misner, Thorne, and Wheeler 1973, chap. 21; Pitts 2004). Since this is generally considered an alternative form of Einstein's field equations rather than a rival theory to GR, whatever empirical support there is for one applies to the other.

For our purposes, the key is that ADM 3+1 decomposes four-dimensional spacetime into three-dimensional spatial components ("hypersurfaces") and one dimension for time. Physicists are mainly interested in this approach because it allows one to set up an "initial value problem": inputting the state of a system and seeing how it evolves. Skipping all the mathematical details,[21] there are components describing a system's spatial geometry and curvature at a point in time. Others connect these temporal slices to one another and describe how they evolve. This is not simply a return to Newtonian space and time, since the hypersurfaces can have odd geometries. It is, however, a textbook version of GR that models change over time as spatial hypersurfaces evolve from one to another. While the data have to conform to certain conditions to make this all work, ADM 3+1 is overall less idealized and requires fewer constraints than cosmic time (Pitts 2004, 6).[22]

This, once again, seems like good news for the presentist. Given that ADM 3+1 and Einstein's field equations are two formulations of the same theory, GR does not *entail* a four-dimensional interpretation or block universe. Why is this alternative not more widely discussed? Again, there are conditions for ADM 3+1 to apply. One is that the systems involved must be "globally hyperbolic," which requires a simple causal relation between temporal slices. Standard GR therefore applies to a wider range of systems. And if global hyperbolicity is a necessary condition for a universe to have a passage of time, then time is again a contingent property that not all physically possible universes have. Time would therefore not be the fundamental aspect of reality that presentists typically take it to be.

One last limitation is that while ADM 3+1 supports the passage of time, it does not reinstate an objective "now" for all observers. As we saw in Section 4.3, the relativity of temporal events in SR is where the problems began for presentism. Observers in different inertial frames have conflicting views about which events have occurred and which are in the future.[23]

With or without absolute simultaneity—an objective "now"—the ADM 3+1 formalism is enough to undermine eternalist claims regarding the block universe. ADM 3+1 shows that relativity is consistent with the passage of time and that GR does not entail eternalism. The correct ontology of time is instead underdetermined by current physics. Other cases of underdetermination in the history of science were resolved with more data.[24] Perhaps future physics will do so here and provide a more presentist-friendly alternative, as Brad Monton has suggested (2006). Sean Carroll is also increasingly sympathetic to proposals that reject Lorentz invariance (Carroll, n.d., loc. 47:35), which was the concept that first brought relativity into the discussion in the philosophy of time.

Physicists agree that GR is not the last word on space and time. Somehow or other, quantum mechanics will have to be squared with GR—an elusive goal to date. Recent proposals for doing so suggest that spacetime is not itself fundamental. Let's briefly consider how this might go.

4.8 A More Fundamental Physics?

Quantum field theory integrates quantum mechanics, field theory, and SR. While it has been an established area of research for decades, there has only been limited success incorporating gravity and GR into this picture. This is partly because gravity in GR is not considered a Newtonian force. The gravitational field is replaced in GR by the geometrical properties of spacetime. Strictly speaking, the sun does not exert a force on the planets in this picture. The mass of the sun instead distorts the spacetime in its vicinity so that the planets travel the straightest path available to them. Effects like gravitational lensing, in which the light of distant galaxies curves around galaxies nearer to us,[25] are likewise due to matter's influence in shaping the geometry of spacetime. Or at least, that is the standard view about GR. Many quantum field theorists now believe this gravity-as-geometry interpretation is not true:

> In the view of quantum theory, it is more illuminating and more useful to say that *there is no spacetime*, there is just the gravitational field. From this point of view, the gravitational field is very much a field like any other field. Einstein's discovery is that the fictitious background spacetime introduced by Newton does not exist. Physical fields and their relations are the only components of reality.
>
> (Rovelli 2008, 367, italics added)

If there is no spacetime and gravity is just another field alongside (quantized versions of) electromagnetism and the like, then GR is much less of a threat to presentism. Without spacetime, there is no block universe.

Instead of denying spacetime outright, other researchers describe it as "emerging" from something more fundamental (Martens 2019; Wüthrich 2019; Huggett, Matsubara, and Wüthrich 2020; Jaksland and Salimkhani forthcoming). Once again ignoring the details, the interesting conclusion for our purposes is that spacetime is taken to be grounded in some other physical reality. This is the sort of research Monton (2006) believes might be more friendly to presentism.

Ellis argues that what lies at the most fundamental level—if there is one—is less important than what emerges from it. "If an effective time emerges at the macro scale, then however that happens, it emerges, and the EBU [evolving block universe] proposal is then good at macro scales, no matter how it relates to a timeless substrate" (Ellis and Goswami 2014, sec. 13.5). Ellis's EBU is one of those near neighbors to presentism mentioned earlier (note 2, this chapter). His point is that a static block universe is undermined even if time is only real on everyday scales. Macroscopic truths about time are still truths.

4.9 One Last Opinion

We have seen in this chapter that the most common way of presenting SR and GR supports a block universe and thereby eternalism. The common way, however, is neither the only nor perhaps the most accurate way of understanding relativity. SR allows for both a dynamical and a geometrical interpretation (Section 4.6). Even if this were not the case, SR is not the best theory of space and time we have. GR is somewhat more accommodating to presentism insofar as it allows for cosmic time and the ADM 3+1 formulation. Future research may undermine the existence of spacetime itself.

I end this chapter with a blatant argument from authority. In case the reader suspects that the contrarians must simply misunderstand something, consider Tim Maudlin, one of the most respected philosophers of physics in the world today:

> [It] is a central aspect of our basic picture of the world that time passes, and that in virtue of that passage things change. And there are no good logical or scientific or philosophical arguments that cast doubt on the passing of time, and there are no impediments to representing, in our present physical theories, that time passes. I draw what ought to be a most uninteresting conclusion, but one that has somehow managed to become philosophically bold: time does pass. Its passage is not an "illusion" or "merely the product of our viewpoint" or "an appearance due to our special mode of perception". Its passage is not a myth. The passing of time may be correlated with, but does not consist in, the positive gradient of entropy in the universe. It is the foundation of our asymmetrical treatment of the initial and final states of the universe. And it is not to be reduced to, or analyzed in terms of, anything else.
>
> (Maudlin 2007, 142)

While Maudlin's view falls short of full-blown presentism, he concludes that physics does not require a static view of time. He might be wrong, but Maudlin does not speak from ignorance when it comes to physics or its philosophical implications.

All this is good news for presentism and for leeway libertarianism. Current physics, by itself, does not entail the existence of a block universe or a fixed future. Libertarian freedom, including the leeway version, is still a live option.

The detour mentioned at the beginning of the chapter has concluded. We will now get back on track with the ways in which a libertarian can deal with physical determinism. The first way appeals to mind-body dualism, but dualism has its own physics-based objections to overcome. These are the subject of the next chapter.

Notes

1 This is another pass at questions taken up in Koperski 2015 (chap. 3).
2 Close cousins to presentism are the *growing block* and *moving spotlight*. The former says that past and present events exist, but not the future (Correia and Rosenkranz 2018; Ellis and Goswami 2014). The latter says that the past and future do exist in some sense, but there is a special place on the timeline that represents the changing present (Cameron 2015). Since our focus here is on the future, the differences between presentism, growing block, and moving spotlight are less important than their common rejection of eternalism.
3 There are eternalist dissenters regarding the passage of time, but not many. See Pooley (2013, 325–26) for a brief discussion.
4 This is the speed of light in a vacuum, something entailed by Maxwell's electromagnetic theory.
5 More precisely, it approximates one. The fact that Earth is a spinning globe means that the room is slowly changing direction.
6 We don't notice this effect because it is hard to detect at speeds less than around ½ *c*.
7 Why is the time coordinate negative? This comes from the spacetime *metric*. Whereas the interval is just the Pythagorean theorem applied to place-time events, the metric has to do with the geometry of spacetime itself. Mathematically, it is a tensor, which is a somewhat advanced device. Other, non-Minkowski spacetimes, like those in general relativity, use different metrics.
8 Some eternalists would dispute that our phenomenology of time amounts to an "illusion" (Qureshi-Hurst 2022, 15–16). Physicists are more comfortable with such language, e.g., Paul Davies: "From the fixed past to the tangible present to the undecided future, it feels as though time flows inexorably on. But that is an illusion" (Davies 2002, 40).
9 What we have here, according to Theodore Sider, "is often (justifiably, I think) considered to be the fatal blow to presentism: that it is inconsistent with special relativity. The notion of the present time that is so crucial to presentism is meaningless within Minkowski spacetime … " (Sider 2001, 42).
10 See Note 2. There are other dynamical or "A-theory" views of time.
11 Timpe calls this type of sourcehood, which is wholly independent of AP, "narrow sourcehood libertarianism" (2012, 147). Wide sourcehood takes AP to be entailed by the sourcehood account, and so the two go together. Narrow sourcehood is a minority view among contemporary libertarians (Taylor Cyr, private correspondence).

12 Katherine Rogers argues that this was roughly Anselm's solution to the problem of divine foreknowledge (2015, 112). The agent at time *t* makes the choice. Eternalism allows God to know that choice for all times prior to *t* since God knows all truths.

13 Two others were mentioned in Note 2, but presentism is more widely held.

14 Ben Page offers a physics-free solution to the problem that some philosophers might find attractive (2022, 86–89). One can understand PAP, he argues, in terms of possible worlds rather than an agent's ability or power. What does it mean that an agent could do otherwise?

One way to see if this condition is met is to think about whether an agent has access to some possible worlds beside our actual one (van Inwagen, 1983, 91). To do this, think of all the closest possible worlds that have the same worldly history and the same laws of nature up until the moment of choice, and then see if the agent acts differently in any of them. If the agent does act differently in some, then we can say that PAP is met. Given what has been said above about eternalism and indeterminism, I cannot see why eternalism would rule out that there are possible worlds in which I act differently. After all, eternalism is not a modal thesis (Page 2022, 87).

If the agent acts differently in other nearby possible worlds under the same circumstances and same laws, then yes, the agent might have done otherwise, just as the leeway libertarian says. There is no need, therefore, to find a way around the block universe.

On the one hand, this seems like a proposal a libertarian would want to explore. There is a possibility at some future time that an agent chooses one course of action rather than another where the alternatives are cashed out in terms of possible worlds. One can therefore hold something like PAP/AP even under eternalism. On the other hand, the contingency described by PAP/AP is a temporal one. There are two different types of possibility being considered here, each with its own logic (modal and temporal) and semantics (possible worlds and right-branching). One can argue that libertarians should have been using a different notion of possibility all along, but leeway proponents do in fact think of theirs in temporal, right-branching terms. Since there is no temporal branching within a possible world, I do not think most leeway theorists will find Page's suggestion attractive.

15 Other antirealists are more radical, rejecting correspondence truth itself. *Social constructivism* takes all purported truth-seeking disciplines—science, history, mathematics, etc.—merely as discourses created by us, like art or traffic laws (Chakravartty 2017, sec. 4.3). To be a "scientific truth," they argue, only requires that the claim be widely accepted by credentialed scientists.

16 This depends in part on one's interpretation of quantum mechanics. The example here only applies to those interpretations that allow for non-local interactions.

17 It has been favorably portrayed more recently by J. S. Bell, at least in terms of pedagogy. The Minkowski approach, he thought, too quickly leads to metaphysical questions about spacetime (Bell 2004, 80).

18 At least prima facie, things tip in favor of the geometrical approach in GR (Huggett, Hoefer, and Read 2023, sec. 4.3), although Pitts argues that the extant

philosophical literature on this is incomplete (Pitts 2019). In SR, neither the interval nor the underlying Minkowski metric changes. In GR, there is instead a metric field that changes due to the presence of matter and energy. On the geometrical approach, spacetime exists apart from both matter and energy. The metric field is considered fundamental, just like the geometry of spacetime. On the dynamical approach, the metric field *arises from* the interaction of matter and energy. The field has no "geometrical significance" in and of itself (Rickles 2007, 739). The distribution of matter/energy and their motions are instead fundamental.

19 Often this local data has to do with the decrease in the average mass-energy density and the temperature of the cosmic microwave background (Torretti 1996, 206–7), but there are other ways to define a fundamental observer (Lockwood 2005, 118–21).

20 Mathematician Kurt Gödel recognized this: "[If] someone asserts that this absolute time [in the FLRW models] is lapsing, he accepts as a consequence that, whether or not an objective lapse of time exists … depends on the particular way in which matter and its motion are arranged in the world. This is not a straightforward contradiction; nevertheless, a philosophical view leading to such consequences can hardly be considered as satisfactory" (Gödel 1949, 562, quoted in Smeenk 2016, 206).

21 See Alcubierre (2008, chap. 2) for more.

22 For a related approach that also argues that GR is compatible with presentism, see Crisp (2008).

23 Pitts shows how absolute simultaneity can be restored within the ADM 3+1 framework (2004, 6–8), but doing so requires adding temporal structure that the field equations themselves do not have. More precisely, he derives Einstein's field equations starting from a particular Lagrangian density that contains a preferred foliation (2004, app. B). Critics will complain that this addition is unnecessary and ad hoc. It has the virtue, however, of being empirically equivalent to Einstein's field equations (Pitts 2004, 8).

24 Theories of combustion based on oxygen and phlogiston could both account for most eighteenth-century observations. The latter eventually gave way under the weight of experimental evidence (Kitcher 1995, 277–79).

25 And which Einstein himself discussed (1936).

Works Cited

Alcubierre, Miguel. 2008. *Introduction to 3+1 Numerical Relativity*. Oxford: Oxford University Press. https://doi.org/10.1093/acprof:oso/9780199205677.001.0001.

Balashov, Yuri, and Michel Janssen. 2003. "Presentism and Relativity." *British Journal for the Philosophy of Science* 54 (2): 327–46.

Bell, John Stewart. 2004. *Speakable and Unspeakable in Quantum Mechanics: Collected Papers on Quantum Philosophy*. Rev. ed. Cambridge: Cambridge University Press.

Bishop, Robert C., and Harald Atmanspacher. 2011. "The Causal Closure of Physics and Free Will." In *The Oxford Handbook of Free Will*, edited by Robert

Kane, 2nd ed., 101–14. Oxford: Oxford University Press. www.oxfordhandbooks.com/view/10.1093/oxfordhb/9780195399691.001.0001/oxfordhb-9780195399691-e-5.

Bourget, David, and David J. Chalmers. 2023. "Philosophers on Philosophy: The 2020 PhilPapers Survey." *Philosophers' Imprint* 23 (1): 1–53. https://doi.org/10.3998/phimp.2109.

Brown, Harvey R. 2005. *Physical Relativity: Space-Time Structure from a Dynamical Perspective*. Oxford: Oxford University Press.

Brown, Harvey R., and James Read. 2021. "The Dynamical Approach to Spacetime Theories." In *The Routledge Companion to Philosophy of Physics*, edited by Eleanor Knox and Alistair Wilson, 70–85. London: Routledge.

Cameron, Ross P. 2015. *The Moving Spotlight: An Essay on Time and Ontology*. Oxford: Oxford University Press.

Carroll, Sean. 2011. "Ten Things Everyone Should Know about Time." *Discover Magazine*. September 1, 2011. www.discovermagazine.com/the-sciences/ten-things-everyone-should-know-about-time.

Carroll, Sean. n.d. "Matt Strassler on Relativity, Fields, and the Language of Reality." *Sean Carroll's Mindscape Podcast*. Accessed March 4, 2024. www.preposterousuniverse.com/podcast/2024/03/04/267-matt-strassler-on-relativity-fields-and-the-language-of-reality/.

Chakravartty, Anjan. 2017. "Scientific Realism." In *Stanford Encyclopedia of Philosophy*, edited by Edward N. Zalta, Summer 2017. Metaphysics Research Lab, Stanford: Stanford University. https://plato.stanford.edu/archives/sum2017/entries/scientific-realism/.

Correia, Fabrice, and Sven Rosenkranz. 2018. *Nothing to Come: A Defence of the Growing Block Theory of Time*. Cham: Springer.

Crisp, Thomas. 2008. "Presentism, Eternalism and Relativity." In *Einstein, Relativity and Absolute Simultaneity*, edited by William Lane Craig and Quentin Smith, 262–78. London: Routledge.

Davies, Paul. 2002. "That Mysterious Flow." *Scientific American* 287 (3): 40–47.

Eagle, Antony. 2021. "Relativity and the A-Theory." In *The Routledge Companion to Philosophy of Physics*, edited by Eleanor Knox and Alastair Wilson, 86–98. London: Routledge.

Einstein, Albert. 1936. "Lens-Like Action of a Star by the Deviation of Light in the Gravitational Field." *Science* 84 (2188): 506–7. https://doi.org/10.1126/science.84.2188.506.

Ellis, George F. R. 2007. "Physics in the Real Universe: Time and Space-Time." In *Relativity and the Dimensionality of the Word*, edited by Vesselin Petkov, 49–79. Berlin: Springer.

Ellis, George F. R. 2024. "Physical Time and Human Time." *Foundations of Physics* 54 (1): 1. https://doi.org/10.1007/s10701-023-00738-2.

Ellis, George F. R., and Rituparno Goswami. 2014. "Spacetime and the Passage of Time." In *Springer Handbook of Spacetime*, edited by Abhay Ashtekar and Vesselin Petkov, 243–64. Springer Handbooks. Berlin: Springer. https://doi.org/10.1007/978-3-642-41992-8.

Ellis, George F. R., and D. R. Matravers. 1995. "General Covariance in General Relativity?" *General Relativity and Gravitation* 27 (7): 777–88. https://doi.org/10.1007/BF02105323.

Gödel, Kurt. 1949. "A Remark about the Relationship between Relativity Theory and Idealistic Philosophy." In *Albert Einstein: Philosopher Scientist*, edited by P. A. Schilpp, 557–62. Chicago: Open Court.

Hoefer, Carl, Nick Huggett, and James Read. 2023. "Absolute and Relational Space and Motion: Classical Theories." In *Stanford Encyclopedia of Philosophy*, edited by Edward N. Zalta and Uri Nodelman, Spring 2023. Metaphysics Research Lab, Stanford: Stanford University. https://plato.stanford.edu/archives/spr2023/entries/spacetime-theories-classical/.

Huggett, Nick, Carl Hoefer, and James Read. 2023. "Absolute and Relational Space and Motion: Post-Newtonian Theories." In *Stanford Encyclopedia of Philosophy*, edited by Edward N. Zalta and Uri Nodelman, Summer 2023. Metaphysics Research Lab, Stanford: Stanford University. https://plato.stanford.edu/archives/sum2023/entries/spacetime-theories/.

Huggett, Nick, Keizo Matsubara, and Christian Wüthrich, eds. 2020. *Beyond Spacetime: The Foundations of Quantum Gravity*. Cambridge: Cambridge University Press. https://doi.org/10.1017/9781108655705.

Ingram, David, and Jonathan Tallant. 2023. "Presentism." In *Stanford Encyclopedia of Philosophy*, edited by Edward N. Zalta and Uri Nodelman, Winter 2023. Metaphysics Research Lab, Stanford University. https://plato.stanford.edu/archives/win2023/entries/presentism/.

Jaksland, Rasmus, and Kian Salimkhani. 2023. "The Many Problems of Spacetime Emergence in Quantum Gravity." *British Journal for the Philosophy of Science*. https://doi.org/10.1086/727052.

Kitcher, Philip. 1995. *The Advancement of Science: Science without Legend, Objectivity without Illusions*. Oxford: Oxford University Press.

Koperski, Jeffrey. 2015. *The Physics of Theism: God, Physics, and the Philosophy of Science*. Chichester: Wiley-Blackwell.

Lockwood, Michael. 2005. *The Labyrinth of Time*. Oxford: Oxford University Press.

Lucas, John. 2008. "The Special Theory and Absolute Simultaneity." In *Einstein, Relativity and Absolute Simultaneity*, edited by Quentin Smith and William Lane Craig, 279–90. London: Routledge.

Martens, Niels C. M. 2019. "The Metaphysics of Emergent Spacetime Theories." *Philosophy Compass* 14 (7). https://doi.org/10.1111/phc3.12596.

Maudlin, Tim. 2002. *Quantum Non-Locality and Relativity*. 2nd ed. Chichester: Wiley-Blackwell.

Maudlin, Tim. 2007. *The Metaphysics within Physics*. Oxford: Oxford University Press.

Maudlin, Tim. 2008. "Non-Local Correlations in Quantum Theory: How the Trick Might Be Done." In *Einstein, Relativity and Absolute Simultaneity*, edited by William Lane Craig and Quentin Smith, 156–79. London: Routledge.

McKenna, Michael. 1997. "Alternative Possibilities and the Failure of the Counterexample Strategy." *Journal of Social Philosophy* 28: 71–85.

Mellor, D. H. 1998. *Real Time II*. London: Routledge.

Miller, Kristie. 2016. "Presentism, Eternalism, and the Growing Block." In *A Companion to the Philosophy of Time*, edited by Adrian Bardon and Heather Dyke, 345–64. Blackwell Companions to Philosophy 52. Chichester: Wiley Blackwell.

Misner, Charles W., Kip S. Thorne, and John Archibald Wheeler. 1973. *Gravitation*. San Francisco: W.H. Freeman.

Monton, Bradley. 2006. "Presentism and Quantum Gravity." In *The Ontology of Spacetime*, edited by Dennis Dieks, 263–80. Boston: Elsevier.

Mozersky, M. Joshua. 2011. "Presentism." In *The Oxford Handbook of Philosophy of Time*, edited by Craig Callender, 1: 122–44. Oxford: Oxford University Press. https://doi.org/10.1093/oxfordhb/9780199298204.003.0006.

Okasha, Samir. 2016. *Philosophy of Science: A Very Short Introduction*. 2nd ed. Oxford: Oxford University Press. https://doi.org/10.1093/actrade/9780198745587.001.0001.

Page, Ben. 2022. "Libertarian Freedom in an Eternalist World?" In *Powers, Time and Free Will*, edited by Anna Marmodoro, Christopher J. Austin, and Andrea Roselli, 83–94. Studies in Epistemology, Logic, Methodology, and Philosophy of Science. Berlin: Springer.

Pais, Abraham. 2005. *Subtle Is the Lord: The Science and the Life of Albert Einstein*. Oxford: Oxford University Press.

Penrose, Roger. 1989. *The Emperor's New Mind*. Oxford: Oxford University Press.

Petkov, Vesselin. 2009. *Relativity and the Nature of Spacetime*. 2nd ed. Berlin: Springer.

Pitts, J. Brian. 2004. "Some Thoughts on Relativity and the Flow of Time: Einstein's Equations Given Absolute Simultaneity." *Chronos (Preprint)*, Proceedings of the Philosophy of Time Society. http://philsci-archive.pitt.edu/2760/.

Pitts, J. Brian. 2019. "Space-Time Constructivism vs. Modal Provincialism: Or, How Special Relativistic Theories Needn't Show Minkowski Chronogeometry." *Studies in History and Philosophy of Science Part B: Studies in History and Philosophy of Modern Physics* 67 (August): 191–98. https://doi.org/10.1016/j.shpsb.2017.05.008.

Pooley, Oliver. 2013. "Relativity, the Open Future, and the Passage of Time." *Proceedings of the Aristotelian Society* (New Series) 113: 321–63.

Qureshi-Hurst, Emily. 2022. *God, Salvation, and the Problem of Spacetime*. Cambridge: Cambridge University Press. https://doi.org/10.1017/9781009269612.

Read, James. 2020. "Geometrical Constructivism and Modal Relationalism: Further Aspects of the Dynamical/Geometrical Debate." *International Studies in the Philosophy of Science* 33 (1): 23–41. https://doi.org/10.1080/02698595.2020.1813530.

Read, James. 2023. *Special Relativity*. Cambridge: Cambridge University Press. https://doi.org/10.1017/9781009300599.

Rickles, Dean. 2007. "Review: Physical Relativity: Space-Time Structure from a Dynamical Perspective." *Mind* 116 (463): 736–40. https://doi.org/10.1093/mind/fzm736.

Rietdijk, C. W. 1966. "A Rigorous Proof of Determinism Derived from the Special Theory of Relativity." *Philosophy of Science* 33 (4): 341–44.

Rogers, Katherin A. 2015. *Freedom and Self-Creation: Anselmian Libertarianism*. Oxford: Oxford University Press.

Rovelli, Carlo. 2008. *Quantum Gravity*. Cambridge Monographs on Mathematical Physics. Cambridge: Cambridge University Press.

Savitt, Steven. 2011. "Time in the Special Theory of Relativity." In *The Oxford Handbook of Philosophy of Time*, edited by Craig Callender, 546–70. Oxford: Oxford University Press.

Sider, Theodore. 2001. *Four-Dimensionalism: An Ontology of Persistence and Time*. Oxford: Oxford University Press. https://doi.org/10.1093/019924443X.001.0001.

Skow, Bradford. 2006. "Review of Harvey R. Brown, *Physical Relativity: Space-Time Structure from a Dynamical Perspective*." *Notre Dame Philosophical Reviews* 2006 (5).

Smeenk, Chris. 2016. "Time in Cosmology." In *A Companion to the Philosophy of Time*, edited by Adrian Bardon and Heather Dyke, 201–19. Blackwell Companions to Philosophy. Vol. 52. Chichester: Wiley Blackwell.

Sommers, Tamler. 2003, March 1. "An Interview with Galen Strawson." *The Believer*. www.thebeliever.net/an-interview-with-galen-strawson/

Timpe, Kevin. 2012. *Free Will: Sourcehood and Its Alternatives*. 2nd ed. New York: Continuum.

Torretti, Roberto. 1996. *Relativity and Geometry*. New York: Dover Publications.

Van Inwagen, Peter. 1983. *An Essay on Free Will*. Oxford: Clarendon Press.

Weyl, Hermann. 1963. *Philosophy of Mathematics and Natural Science*. New York: Atheneum.

Wüthrich, Christian. 2019. "The Emergence of Space and Time." In *The Routledge Handbook of Emergence*, edited by Sophie Gibb, Robin Findlay Hendry, and Tom Lancaster, 315–26. Routledge Handbooks in Philosophy. London: Routledge.

Zimmerman, Dean. 2011. "Presentism and the Space-Time Manifold." In *The Oxford Handbook of Philosophy of Time*, edited by Craig Callender, 163–244. Oxford Handbooks. Oxford: Oxford University Press.

5 Physics vs. Dualism

5.1 Two Arguments from Physics

Chapter 3 ended with two ways that libertarians might escape physical determinism: mind-body dualism and emergence. The latter will be discussed starting in Chapter 7. The former relies on the idea that the laws of nature apply only to physical systems. If the capacity for making free decisions resides in a non-physical aspect of human persons, often called a "mind" or "soul," such a thing is beyond the scope of the laws of nature.

Dualism is, of course, contentious. The sole question here is whether dualism conflicts with physics. (Similar issues confronting emergence will be dealt with in later chapters.) We will consider two physics-based objections.[1]

The first is the *causal closure of the physical* (CCP). Closure says that physical events have only physical causes. CCP would rule out interactive dualism in that bodily movements are physical, and so the cause of those movements must also be physical. No physical event is part of a causal chain that traces back to a non-physical entity.

The second is the *conservation of energy*.[2] For a non-physical mind to causally influence the brain, the objection goes, it would have to create energy. But conservation of energy, a fundamental law of nature, forbids the creation or destruction of energy. Hence, even if there are minds, they cannot interact with brains without violating a law of nature. That's bad. When philosophical positions run aground of well-established laws, those positions quickly become untenable.

I argue that the conventional wisdom regarding CCP is false. Rather than an entailment of physics, CCP is an aspiration of a physicalist approach to science and metaphysics. It cannot be treated as a trump card against all critics of reductionism. The conservation objection is in some ways more interesting but is based on an oversimplification of conservation laws. Once the details are filled in, conservation is less daunting than commonly portrayed. In the end, interactive dualism—henceforth just

DOI: 10.4324/9781003681199-5

"dualism"—faces no insurmountable challenges from physics. It offers one avenue for libertarian free will to be possible in a material world described by deterministic physical laws.

5.2 The Causal Closure of Physics

Let's say a bit more about closure. It starts with the idea that every physical effect having a cause has a sufficient physical cause:

> When you trace the causal history of any physical effect—that is, of anything physical that has a cause—you will never need to appeal to anything non-physical. The physical universe contains within itself the resources for a full causal explanation of any of its (caused) elements, and in this sense is "complete".
>
> (Robb and Heil 2021, sec. 2.4)

Robb and Heil show how "causally complete" might be distinguished from "causally closed," but most use the two terms interchangeably. Whatever distinctions can be made, they are not important for our purposes. They go on:

> The point applies, then, to whatever might occur to or within our bodies. Any instance of bodily behavior has a sufficient physical cause, which itself has a sufficient physical cause, and so on. In tracing the causal history of what we do, we need never appeal to anything non-physical.
>
> (Robb and Heil 2021, sec. 2.4)

Causal chains involving human bodies ultimately involve physical events and nothing else. This is the sense in which the physical is causally "closed."

For all the talk about the conflict between determinism and free will in previous chapters, CCP is the more significant threat to libertarianism. Under CCP, mental events cannot influence the behavior of physical bodies unless they can be reduced to physical events. That, however, would bring such events, including choices, within the scope of the laws of nature—the very thing that libertarianism is trying to escape. There is a clear conflict, then, between the CCP and both libertarianism and dualism. The question remains, why believe CCP is true?

One answer is given by Sean Carroll, a free-will skeptic from Chapter 1, in discussing souls and life after death. He argues that an immaterial mind/soul interacting with material bodies would violate a law of nature, namely the Dirac equation for the behavior of electrons (and other spin ½ particles). If minds are active and present,

> There needs to be a new term (at minimum) on the right [of the Dirac equation], representing how the soul interacts with electrons. (If that term doesn't exist, electrons will just go on their way as if there weren't any soul at all, and then what's the point?) So any respectable scientist who took this idea seriously would be asking—what form does that interaction take? Is it local in spacetime? Does the soul respect gauge invariance and Lorentz invariance? Does the soul have a Hamiltonian? Do the interactions preserve unitarity and conservation of information?
>
> (Carroll 2011)

Not only is no answer forthcoming, says Carroll, but the ramifications of believing in souls would be catastrophic to science:

> Nobody ever asks these questions out loud, possibly because of how silly they sound. Once you start asking them, the choice you are faced with becomes clear: either overthrow everything we think we have learned about modern physics, or distrust the stew of religious accounts/unreliable testimony/wishful thinking that makes people believe in the possibility of life after death. It's not a difficult decision, as scientific theory-choice goes.
>
> (Carroll 2011)

Carroll doesn't say here why the activity of minds would "overthrow everything we think we have learned about modern physics." Elsewhere he is explicit that the particles in the Standard Model respond only to local, physical causes (Carroll 2016a). To reject closure at this level would involve the rejection of a deep truth of physics.[3]

Some physicalists consider closure to be obviously true given that we know the constituent parts of things (Hossenfelder 2019, 90–91). That falls somewhat short of a rigorous argument, however. In fact, there were no such arguments, according to philosopher David Papineau, until he set out to provide one (2002, 45). First, he makes it clear why CCP rules out interactive dualism. If closure holds, "then there is no room left for anything non-physical to make a difference to physical effects, so anything that does make such a difference must itself be physical" (Papineau 2000, 179). While he believes that, strictly speaking, CCP is compatible with the existence of minds, they cannot have any causal influence on brains. Interactive dualism is therefore false, although epiphenomenalism[4] would be possible (Papineau 2009, 54).

Papineau takes physics to be unique in its closure. In contrast,

> The biological realm … is not causally closed in this sense, since biological effects often have non-biological causes, as when the impact of a meteorite precipitated the extinction of the dinosaurs. Again,

> meteorology is not causally closed: the burning of carbon fuels—a non-meteorological event—is causing global warming. Nor, importantly, is the mental realm causally closed: a mental pain can be caused by sitting on a physical drawing pin, or a train of thought can be interrupted by a loud noise.
>
> (Papineau 2009, 54)

But since physics is causally closed, mental causes—if there are any—cannot produce physical change. Only the physical, says Papineau, has causal *oomph*.

Before getting to the evidence for CCP, let's consider two quick points. First, how does CCP apply to indeterministic quantum events? Aren't indeterministic events by definition uncaused? If so, aren't they a counterexample to CCP? To the last question, the answer is "no." The definition of CCP offered by Robb and Heil allows for uncaused events.[5] *If* a physical event has a cause, they say, then that cause must be physical. Moreover, indeterministic quantum events do have necessary conditions, which are sometimes described as causes. The outcome of the Schrödinger's cat example (Section 3.4) depends on the experimental setup and the dynamics described by Schrödinger's equation. It is not "uncaused" in the sense of being wholly spontaneous. There just isn't a sufficient cause determining the specific outcome of dead-cat or live-cat.[6] Quantum mechanics is therefore not a threat to CCP.

Second, some might rightly wonder what exactly is meant by "physical" in CCP. There is some sheepishness about this (Papineau 2009, 57–58). Does it mean the ontology of current physics? If so, then CCP could be refuted by future discoveries. Physicalists surely don't want that. Does "physical" instead mean an ideal, completed physics? That seems problematic too. Who knows what the ontology of such a system will be? Without such knowledge, "physical" becomes a label for some unknown, yet-to-be-determined system—a promissory note that future physicists will one day tell us what our ontology is. Neither option is particularly attractive, which explains why—somewhat ironically—there is no consensus among physicalists on what it is to be physical. Suffice it to say that it's something very small, says Papineau (2009, 58), although one wonders what astrophysicists think of that. (Galaxies surely count as physical.)

In any case, why think that CCP is true?

5.3 Assessing the Arguments for CCP

Let's now assess the three main arguments offered in support of CCP: evidence from physiology, appeals to parsimony, and the alleged conflict with conservation laws.

5.3.1 Physiology

In the age of MRI machines and other advanced diagnostic equipment, physiology has clearly earned its place among the mature sciences. With decades of successful research now in hand, there is no sign of any force or causal influence beyond the physical (Montero 2003, sec. 5; Papineau 2002, 255–56; Papineau 2015, 372–74; Sapolsky 2023, chap. 3). Many philosophers and neurophysiologists believe that the operation of the brain explains everything in need of explanation when it comes to mental phenomena. If there were something non-physical at work, we are told, scientists surely would have discovered some evidence of it by now:

> Were *Completeness* false owing to the routine intervention of souls, physiology would have discovered anomalies within the human body, events that have no physical explanation. But a century of increasingly detailed investigation has revealed no such anomalies. It appears as if the causes operating within our bodies are entirely physical.
>
> (Robb and Heil 2021, sec. 2.4)

More succinctly, "there are no nonphysical causes of physical phenomena because after extensive investigation, none have been found" (Montero 2003, 182). For many, the overwhelming empirical evidence for CCP is the final word on the subject. Those opposed do not appreciate the relevant science.

There are several things to note about this. First, the claim here is about physiology, not physics. Why is an argument for CCP appealing to biology? In practice, the P in CCP often gets expanded to include all the natural sciences. Papineau is explicit about this, calling it "broad physical causal completeness" (2008). Not all physicalists approve of this expansion. Jaegwon Kim, for example, believes that only fundamental physics—whatever it amounts to—is causally closed. Extending closure to include anything more would be a rejection of physicalism itself (Kim 1998, 40). Papineau's broad physical causal completeness also seems to conflict with a passage quoted earlier: "The biological realm, for example, is not causally closed in this sense, since biological effects often have non-biological causes, as when the impact of a meteorite precipitated the extinction of the dinosaurs" (Papineau 2009, 54). If biology is not causally closed, then neither is physiology. Perhaps this was just a poorly chosen example on Papineau's part.

Second, nothing Papineau raised about physiology constitutes direct evidence for CCP. The data are part of an inference to the best explanation. Roughly, (i) if some sort of non-physical entity influenced the brain, there would be evidence for it. But (ii) there is no such evidence. Therefore, it is most likely that (iii) the only causal interactions in the brain are physical. While there is nothing wrong with inference to the best explanation, it should not be conflated with evidence, as Papineau seems to (2009, 54).

Other philosophers are less sanguine about the lack of direct evidence for closure. Nancy Cartwright, for example:

> the claims of closure for physics—whether self-closure or even grander claims that physics can account for every feature of the empirical world that can rightly be called physical, or that physics can account for every feature of the empirical world—are all claims that call for empirical evidence. But we simply do not have the right kinds of evidence to have confidence in any answer. Any answers now, one way or the other, are sheer metaphysics ...
>
> (Cartwright 2011, 618)

Cartwright is a non-reductive physicalist,[7] who denies CCP. She believes the inferential gap between the physiological data and CCP is much larger than its proponents aver, requiring too many "in principle" qualifiers and promissory notes.

Robert Bishop suggests that the lack of evidence against CCP is due in part to the fact that no one has been looking for it (2010, 209). Dualism and other non-reductive programs have not been refuted so much as ignored. Physiology, endocrinology, and neurobiology typically explore only the physical mechanisms responsible for their respective data. They never consider non-physical mental causes. If some observations remain unexplained, researchers either assume that physical mechanisms are somehow or other sufficient for the observed effect (Lowe 2010, 74–75) or that there currently is no explanation. Non-physical causes are not considered. One should not pretend, then, that science has resolved the matter. As Bishop points out: "In this light, Montero and Papineau's construal of physiological research is more like, 'We aren't looking for it, didn't find it, so it's not there!' " (2010, 210).[8]

Overall, the argument from medical research is less conclusive than physicalists typically say. Perhaps there is a way to steelman the claim.[9]

5.3.2 Parsimony

Physicalists often appeal to parsimony. Consider how Papineau begins the following passage by talking about evidence:

> [During] the late nineteenth and the twentieth centuries an increasing number of scientists have come to doubt the existence of vital and mental forces. The most significant evidence seems to have come

> directly from physiology and molecular biology, rather than from physics.
>
> (2009, 54)

Again, this seems like a straightforward appeal to the evidence. What follows is slightly different:

> Over the last hundred and fifty years a great deal has come to be known about the workings of biological systems (including brains), and there has been no indication that anything other than basic physical forces is *needed to account* for their operation.
>
> (Papineau 2009, 54, italics added)

This passage is not merely about evidence but rather competing explanations. If an entity is not needed to explain some phenomenon, that entity is superfluous. Ockham's Razor dictates that one should choose the simpler explanation.[10] The argument seems to be something like this:

1. If there were non-physical causes, they would produce anomalies from a purely physical point of view (effects one would not expect if all the physical causes were known).
2. Physiology has found no such anomalies.
3. Hence, there is no need to posit the existence of non-physical causes.
4. Hence, parsimony favors physicalism.

There is no question that physicalism is more parsimonious than most rival positions in metaphysics. And let's agree that parsimony is a virtue in both science and philosophy. It is, however, not the only one. Anaximenes argued that everything is made of air, which is parsimonious indeed. Why then have so few heard of him? Because in seeking a comprehensive understanding of reality, we draw on other explanatory virtues such as empirical adequacy, explanatory power, and fit with what we already know. As with parsimony, none of these provides a sufficient condition for a good explanation, so the more the better. Note that some virtues are often in tension with others. Solipsism—only my mind exists—is parsimonious. The question is whether it can explain everything that needs to be explained concerning reality. The same goes for physicalism. Can it explain qualia? Can it explain willing? Those are some of the key questions in the philosophy of mind these days.

While physicalism may be more parsimonious, this virtue alone does not establish CCP or rule out dualism, especially if the latter offers greater explanatory power regarding consciousness or agency.

5.3.3 Conservation of Energy

The most robust argument in favor of CCP involves conservation laws. The reasoning is that any system that violates CCP would also violate the conservation of energy. Since conservation is a fundamental law of nature, non-physicalist systems would conflict with physics. Most of the examples in the literature focus on dualism:

> It seems impossible to make substance dualism consistent with modern physics. Physics says that the amount of matter/energy in the universe is constant; but substance dualism seems to imply that there is another kind of energy, mental energy or spiritual energy, that is not fixed by physics. So if substance dualism is true then it seems that one of the most fundamental laws of physics, the law of conservation, must be false.
>
> (Searle 2004, 42)[11]

> Soul–body interaction would have to change the amount of energy in the physical universe. When souls act, new energy would appear in, say, the brain. When souls are acted upon, some quantity of energy in the brain would vanish. But either scenario would contravene established conservation laws, which permit only the conversion and redistribution of energy (or mass–energy) within the physical universe, not its addition or subtraction.
>
> (Robb and Heil 2021, sec. 2.3)

> Dualism violates conservation of energy. If immaterial mind could move matter, then it would create energy; and if matter were to act on immaterial mind, then energy would disappear. In either case energy—a property of all—and only concrete things would fail to be conserved.
>
> (Bunge 1980, 17, quoted in Cucu 2022, 92)

This is no minor conceptual tension. According to Bunge, if we were to believe in dualism, "physics, chemistry, biology, and economics would collapse" (1980, 17). The stakes, it would seem, are quite high indeed.

The argument from conservation undermines the reason for considering dualism in the context of free will. The hope was that if free will is a capacity of an immaterial mind, and minds are not subject to the laws of nature, then the immaterial mind would lie beyond the reach of deterministic dynamical laws. That may be so, but if dualism would violate a different but equally fundamental law, then nothing has been gained.

Either way, dualism seems to conflict with physics, which Daniel Dennett calls its "inescapable and fatal flaw" (1991, 35).

Papineau ties conservation directly to CCP. The only way for a non-physical cause to conserve energy would be to store and release energy in a deterministic way (2009, 57). To understand what he has in mind here, let's start with a distinction. There are two types of dynamical systems in physics: conservative and dissipative. The latter involve friction, drag, and electrical resistance in which energy is lost over time. Systems at the level of the Standard Model, in contrast, are all conservative, meaning there are no such losses, and so conservation of energy applies. Since all macroscopic objects are composed of particles described by the Standard Model, physicists typically believe that dissipative systems can be reduced to conservative ones. Dissipation, then, is a macroscopic mere appearance. The only truly fundamental forces in nature preserve the conservation of energy. If so, then for a non-physical force to influence a physical system, that force must also be subject to conservation laws. A non-physical system for storing and releasing energy would therefore be required. While Papineau does not argue against such a possibility, he notes that scientists around the turn of the twentieth century simply gave up on the idea of such systems (2009, 57). Since there is no obvious way that non-physicalists can meet the conservation requirements, the conservation of energy supports CCP.

Although this argument is a bit loose, let's grant the conclusion: If conservation of energy is a fundamental law of nature with no exceptions, that supports CCP. Physics trumps metaphysics.

Cucu and Pitts analyze several replies a dualist might make, the last of which they call the "Optimal Dualist Response" (2019, sec. 6).[12] Their main idea is that conservation laws do admit exceptions. Conservation laws are conditional, not absolute. This surprises many physicists who are often taught that the laws of nature are absolute and unconditional, with conservation of energy being one of those laws. There can't be a time when, say, the law of universal gravitation does not apply—every bit of matter in the universe attracts every other bit at all times without exceptions or conditions.

But this is not the case for conservation laws. All conservation laws are conditional—and, to the point, it is physically possible for those conditions not to be met. Where do we see this conditionality?

5.3.3.1 General Relativity

Einstein envisioned a static universe, even though the field equations of general relativity seemed to show otherwise. (He famously added the cosmological constant to correct what he then thought of as a flaw.) In such a universe, energy would be globally conserved. But, energy is not conserved in an expanding universe:

> I said that energy is conserved, but that is only approximately correct. It would be entirely correct for a universe in which space does not change with time. But we know that in our universe space expands, and this expansion results in a violation of energy conservation.
>
> (Hossenfelder 2018)

> The conservation-of-energy principle serves us well in all sciences except cosmology. In bound regions that do not expand with the universe … , we can trace the cascade and interplay of energy in its multitudinous forms and claim that it is conserved. But in the universe as a whole it is not conserved. The total energy decreases in an expanding universe and increases in a contracting universe. Where does energy go in an expanding universe? … The answer is nowhere, because in the cosmos, energy is not conserved.
>
> (Harrison 2000, 349)

Carroll shows how the total energy[13] contribution of matter, radiation, and vacuum energy in an expanding universe "is clearly not conserved… This is upsetting, since conservation of energy is one of the more cherished principles of physics" (2019, 120).

While this exception to conservation of energy is well known to cosmologists, it is seldom discussed at the undergraduate level. The technical details of conservation in relativity depend on special mathematical tools called *Killing vectors*, which identify the symmetries of spacetime—essentially the ways that space and time can be "shifted" or "rotated" without changing the underlying geometry (Crowell 2022, sec. 7.1). There is no expansion or contraction of the distance between the points related by such a vector. Killing vectors are required in a space in order for energy to be defined. According to the special theory of relativity, there is an infinite number of Killing vectors, so that energy can be defined and conserved. For most universes described by general relativity, including our own, things are different. There are no Killing vectors. As Tim Maudlin et al. conclude, "There is therefore no reason … to expect any principle of exact global conservation of energy to obtain. The conclusion is admittedly somewhat disconcerting, but there it is" (Maudlin, Okon, and Sudarsky 2020, 70).[14] In short, conservation of energy is conditional, and those conditions do not generally hold in curved spacetimes.

In terms of the applicability of conservation, the key word is "conditional." Conservation of energy is not an absolute law of nature that applies in all circumstances. While it fails in general relativity except for very special cases, it holds for the rest of physics. Hossenfelder says that the expansion of the universe "violates" conservation, but it would be better to say that the conditions required for conservation of energy fail to obtain. By analogy, driving on the right side of the road in Michigan does not violate British traffic laws. Those laws don't apply in the US.[15]

How does this apply to dualism? We'll get to that in the next section. The point here is that conservation laws are not absolute.

5.3.3.2 Closed Systems and Lagrangian Mechanics

A note to the reader: this section is slightly more mathematical than the others. If you want to skip to the conclusion, no great harm will be done.

A simpler and better-known condition is that conservation of energy applies only within closed or isolated systems—where "closed" means that no particles enter or leave the system; "isolated" means no matter or energy is exchanged with the system's environment. While it is possible for an open system to conserve energy, it will not do so in general. Again, talk of "violation" is misplaced. The law is conditional, and open systems do not meet those conditions. This is why the oft-repeated "energy cannot be created or destroyed"—which comes from the first law of thermodynamics—is misguided when applied to brains and bodies. Neither constitute closed systems.

Cucu and Pitts put this in more rigorous terms by appealing to *Lagrangian mechanics* (2019, 107–10).[16] The terminology of "open" and "closed" systems is set aside in favor of a mathematical function called a *Lagrangian*, which describes the energy of a system, specifically the kinetic energy minus the potential energy. The kinetic energy might be the familiar $\frac{1}{2}mv^2$, where v is the velocity and m is the mass of the object in question. The potential energy is more system-specific. For a spring oscillating along a surface, it would be $\frac{1}{2}kx^2$, where k describes the stiffness of the spring and x is the distance the weight moves.[17] For the spring, then, the Lagrangian $L = \frac{1}{2}mv^2 - \frac{1}{2}kx^2$.

Notice that there is no variable t for time in this equation. L is not a function of time, which means that L cannot change over time. This is important. According to a famous theorem by Emmy Noether ([1918] 1971), conservation of energy depends on the Lagrangian. If the Lagrangian for a given system remains constant over time, then energy is conserved. The ideal spring meets this criterion. L does not change over time as the spring oscillates. If the Lagrangian is a function of time, on the other hand, the system may or may not conserve energy.[18]

Noether's proof can be strengthened. For many systems, the relation is biconditional: The Lagrangian changes over time if and only if energy is not conserved (José and Saletan 2006, 251). Consider the moving spring again. Suppose the one-time family cat, Arthur (who I do not miss), took a swipe at the moving weight. In the swiping case, the Lagrangian would no longer be L. The correct function would have to account for Arthur's interference on the spring. Such a function would explicitly depend on time, taking on one form when the cat is active and another when it is not. According to the biconditional version of Noether's Theorem, this shows that conservation of energy does not apply for this system.[19] The crucial point for the dualism debate is this conditionality shows, once again, that conservation laws are not absolute.

Now let's apply this to the interaction between a non-physical mind and the physical brain. For the dualist, the mind is like the cat. If one's brain is the system under consideration, that system is influenced in different ways by a mind at different points in time. A Lagrangian describing the brain would include that influence (even if such a function is too mathematically complex to calculate). This means, once again, that the Lagrangian would be time dependent: it would differ as the mind's influence on the brain changes over time. By Noether's Theorem, conservation of energy does not apply to brains and so is not violated. A parallel argument can be made for conservation of momentum, but the conclusion is the same:

> [The] action of a non-physical mind on the brain, assuming that the mind does not act in exactly that same way at every moment in time and every point in space throughout the entire universe (which seems a safe assumption), does indeed entail non-conservation of energy (by singling out some times as those when the mind acts on the brain) and non-conservation of momentum (by singling out some places as those where the mind acts, presumably within the brain only).
>
> (Cucu and Pitts 2019, 104)

Conservation therefore does not pose a threat to interactive dualism.

Of course, physicalists do not believe in minds. But let's not pretend that physics or the preferences of physicists have resolved that philosophical question. Conservation laws are conditional, and if dualism is true, human brains are exceptions to those conditions.

5.3.3.3 Objection

Physicalists have a response to the argument that conservation laws are conditional. They claim that apparent violations of conservation can always be "fixed" by expanding our view of the system. I'll examine this objection and show why it fails.

In response, physicalists argue that the failure of conservation is merely prima facie. Some forces, like friction, seem to make energy disappear. Imagine a ball rolling on a rough surface. Friction makes it slow down—where did that kinetic energy go? Into heat in the surface. If we account for that heat, energy is conserved. Physicalists argue this pattern applies everywhere. The energy is always there. We just need to look more broadly. This is the case, they argue, for all so-called "dissipative forces" including electrical resistance and aeronautical drag. Energy is never lost; it is instead changed into some other form of energy, as Papineau argues:

> Helmholtz's formulation of the conservation of energy hinged on the assumption that friction and other dissipative forces are nonfundamental forces, macroscopic manifestations of processes involving more fundamental conservative forces. For it is only if we see macroscopic forces like friction as reducing to fundamental conservative forces that we can uphold the universal conservation of energy. Now, this point can be viewed as providing inductive support for the general thesis that *all* apparently special forces will reduce to a small stock of fundamental forces. The special forces that have been quantitatively analyzed, like friction, turn out to reduce to more fundamental conservative forces. So this provides inductive reason to conclude that any other apparently special forces, like muscular forces, or vital forces, or mental forces, will similarly reduce.
>
> (Papineau 2015, 370)

The idea here is that in a sense, the force of friction isn't real. Nonfundamental, dissipative forces can always be reduced to fundamental forces that are not dissipative and therefore conserve energy.

For this idea to work, one must always be able to redraw the boundaries of a system to change an open system into a closed one. By analogy, while magicians might seem to violate conservation and other laws of physics, if we expand our view to include their hidden apparatus, we see that they do not. The physicalist claims we can always do something like this to show that conservation of energy is never violated. If Arthur the cat, for example, is considered part of a composite cat-spring mechanism, then the cat isn't interfering with the system. It is part of the system. Although no one could write the Lagrangian for the cat-spring, one exists in principle that is not time dependent and so conserved energy. This is the sense in which a dissipative force can be "reduced to" a conservative one, as Papineau says. Once we grasp this truth about dissipative systems, it seems unlikely that conservation of energy ever actually fails.

Is this the case? For many free-will skeptics, it's obvious. Bodies are composed of particles in the Standard Model. The Standard Model does not include dissipative forces. So somehow or other, dissipative systems must reduce to non-dissipative ones.

Such reductionist arguments are often borne along by the qualifiers "in principle" or "somehow must."[20] There is nothing here like the clear and precise conditions provided by Noether. More importantly, there are decisive counterexamples where the strategy of rejiggering the borders cannot work, no matter how clever we are about redrawing those boundaries. First, we've already considered why conservation of energy does not apply to the universe as a whole. The issue there has nothing to do with dissipation, and so the redrawing of boundaries is irrelevant. Second, there is the GRW (Ghirardi-Rimini-Weber) interpretation of quantum mechanics. The details don't matter here—what matters is that in this interpretation, energy genuinely disappears during quantum measurements, and no amount of boundary-redrawing can fix this.[21] Carroll and Lodman (2021) go further, arguing that this is the case for all collapse interpretations of quantum mechanics, including the Copenhagen interpretation. Again, no redrawing of boundaries will reestablish conservation of energy in that case. Third, there are reasons to believe that all realistic systems are open, not closed, and that conservation only applies to idealized systems (Rivat 2024).

These failures show that the "principle of the conservation of energy is not sacrosanct" (Butterfield 1998, 146–47). Maudlin et al. are bolder: "It seems that we must accept the fact that neither conservation laws nor Einstein's equations hold in the world we actually experience" (Maudlin, Okon, and Sudarsky 2020, 80). Given the conditional nature of conservation laws, the claim that conservation constitutes dualism's "fatal flaw" (Dennett 1991, 35) has not aged well.[22]

Having examined the primary arguments for CCP—appeals to physiology, parsimony, and conservation laws—we find each comes up short. The physiological argument rests on an inference to the best explanation where counterevidence hasn't been actively sought, parsimony must be weighed against other explanatory virtues, and the crucial conservation of energy argument fails because conservation laws are conditional and likely don't apply to brain-mind interactions if dualism is true.

5.4 Whither CCP?

Having examined the key arguments for CCP, we can now assess its overall status and implications for dualism. Is there any role that it might play? Kim believes there are pragmatic reasons to accept CCP, arguing that it is essential for the long-run success of physics. If physics is not closed, then "to explain some physical events you must go outside the physical realm and appeal to nonphysical causal agents and laws governing their behavior! Complete physics would in principle be impossible, even as an idealized goal" (Kim 1996, 147). But is the goal of a complete physics needed as a spur to research? As we will see in Chapter 6, many who think of themselves as non-reductive physicalists or emergentists have mostly given up on the idea of a complete, harmonized physics.

Still, there isn't much harm in employing ideal physics in an aspirational sense. Surely researchers can use CCP as a provisional methodological principle if they find it useful. Experimentalists working with isolated physical systems can accept CCP unless they have some reason to believe otherwise. Non-physical causes, if there are any, are unlikely to interfere with the goings on in, say, a laser optics lab or a particle accelerator. But let's be clear: That sort of research has no more to say about the existence of non-physical causes than it does about the existence of Detroit. It's simply not looking into the question of non-physical influences. Bishop puts it this way:

> Physics does not imply its own closure... Rather, CoP [i.e., closure of the physical] is a metaphysical doctrine. Indeed, physics tells us what happens when particular forces are taken into account, but nothing about what happens when influences unaccounted for by physics are present.
>
> (Bishop 2011, 606)

Jessica Wilson says much the same: "*Physical Causal Closure* is not a principle of contemporary physics (though no doubt many physicists, and scientists more generally, believe it). The acceptance of this principle is rather a constraint on physicalist theorizing ..." (2021, 124).

Consider, in contrast, the principle that nothing with mass can be accelerated to the speed of light. This is a direct consequence of the mathematics of special relativity. No such derivation exists for CCP. It is not found in the field equations of general relativity, nor is it a theorem of the Standard Model. It is, rather, a philosophical constraint imposed upon science, motivated by a prior commitment to physicalism. Physics itself is at best neutral on the matter. Portraying CCP as somehow entailed by science is a misrepresentation. Papineau, somewhat grudgingly, seems to agree:

> Of course, as with all empirical matters, there is nothing certain here. There is no knock-down argument for the completeness of physics. You could in principle accept the rest of modern physical theory, and yet continue to insist on special mental forces, which operate in as yet undetected ways in the interstices of intelligent brains.
>
> (Papineau 2002, 256)

It would be obstinate, he thinks, to reject CCP, but physics would survive without it. Barbara Montero concurs:

> [Some] may still have a vague nagging feeling that rejecting CC is somehow being antiscientific. But rejecting CC pits one against science no more than accepting the incompleteness theorem pits one against mathematics. Neither the particular argument nor the vague nagging feeling that rejecting CC is antiscientific is well grounded. Science can be as successful as you like and CC could still be false.
>
> (Montero 2003, 180)

All this shows that the stakes are not nearly as high as some have thought. Bunge claimed that the rejection of physicalism and CCP would entail the violation of conservation of energy and thereby the collapse of the natural sciences (1980, 17)! Papineau paints a much less dire picture.

Free-will skeptics like Hossenfelder instead assert that CCP is obvious. She claims that brains, like all macroscopic objects, are composed of particles, and that one can predict the behavior of the whole from its constituent parts (Hossenfelder 2020). Atomism thus ends the debate. Dualists correctly note that this begs the question. One can agree about the composition of brains and yet deny that composition is the whole story regarding human persons. Better-informed physicalists like Papineau understand that science does not prove CCP. Closure is an assumption of reductive physicalism, not an independent argument in its favor.

There is one final reason physicalists might hesitate to invoke CCP. Many particle physicists take causation itself to be an emergent phenomenon that plays no role in the Standard Model (Carroll 2016b, 375). They do not believe causation is real, at least not in the way that photons and electrons are. This echoes a famous argument by Bertrand Russell that causation in physics has been replaced by the notion of laws of nature (1913). While Russell's conclusion is overstated,[23] it is at least a tad embarrassing to declare that physics is "causally closed" and then have physicists like Carroll tell you that causation doesn't apply to fundamental physics.

I have not tried to answer every objection to dualism here. There is still, for example, the question of interaction: how can an immaterial entity causally interact with a physical one?[24] The influence of minds is mysterious, at least compared to contact forces like my hand pushing a glass across a table. Compared to fields, on the other hand, things are more on a par. Fields interact with particles and with each other, and any number of them are also able to occupy the *same* space at the same time. That's hard to conceptualize too. At some point, one simply has to say that's just the way it is.

In the end, dualism successfully provides a means for libertarian action in an otherwise physical world, or at least there is nothing in physics itself that forbids it. There is no conflict between dualism and science. The conflict is between dualism and physicalism.

Notes

1 There are many more objections, but the rest are philosophical. I leave that debate to others. For a recent defense of dualism, see Rickabaugh and Moreland (2023). For a defense of physicalism, see Gillett and Loewer (2001).
2 Arguments involving conservation of momentum are sufficiently similar that they do not warrant their own treatment.
3 See Pitts (2020, 1940) for more on Carroll's argument, which Pitts believes begs the question. A dualist could say that the Dirac equation holds in particle accelerators but is an incomplete description of human persons. This doesn't seem to be much different from saying that Einstein's field equations describe the solar system but have to be adjusted for dark matter when describing galaxies.
4 Epiphenomenalism says that while brains produce mental phenomena, mental states are only ever effects, never causes (Robinson 2023). CCP also rules out any of the stronger forms of emergence to be discussed in later chapters. Higher-level, emergent causes would at best be redundant if they are bringing about physical changes. The only real causes, according to CCP, reside at the most fundamental level of physics.
5 Cf. Papineau (2009, 59).
6 Not everyone is happy with this characterization. Philosopher E. J. Lowe argues that one cannot easily reformulate CCP so that it allows for probabilistic causes (2003, 143–44).
7 What that is will be made clearer in Chapter 6. In short, non-reductive physicalists agree that the only things that exist are physical—composed of subatomic particles, fields, etc. They are pessimistic, however, about our ability to explain all higher-level phenomena in terms of more fundamental theories. Textbook science will never be fully unified.
8 Bishop et al. make a similar point elsewhere:
Many philosophers of mind and metaphysicians over the years have argued that CoP is well supported by science and that the burden of proof is on those who would argue against it… As CoP is supposed to be an empirical claim, we can start by asking what would constitute evidence of a violation of CoP? Loewer gives us a clue: some deviation or gaps in the laws… [But] just consider the vagueness of the criterion. No one ever specifies what would count as a genuine violation or gap in the laws. Typically, any proposed violation or gap is dismissed as "the effect of an unknown physical cause." If one is already convinced of the truth of CoP, then that is the stock response one would have to give (Bishop, Silberstein, and Pexton 2022, 292).
9 A steelman argument is the opposite of a strawman: it strengthens the opposing side's case rather than caricaturing it.

10 There are different versions Ockham's razor and simplicity one might employ. See Fitzpatrick (2013) for an introduction.
11 While Searle's characterization is a bit loose, Cucu and Pitts show how it can be strengthened (2019, 97).
12 Pitts lays out the technical details in a series of papers (Pitts 2019, 2020, 2021).
13 More precisely, the energy density over which an integral can be taken.
14 See also Carroll (2010).
15 It is worth noting that philosopher Robin Collins (2008, sec. IV; 2011, 125–30) and physicist Ulrich Mohrhoff (1997, sec. 3) raised the issue of cosmology and conservation some years ago in the context of objections to mind-body dualism. While Pitts disagrees about their characterization of general relativity (2020, sec. 6), the important point for our purposes is their agreement that conservation of energy has conditions for it to hold.
16 Their version is more precise (and complex) as it applies to fields. The same basic idea can be illustrated with the simplified examples provided here.
17 This Lagrangian is for an "ideal" spring, one without friction or gravity. The example could be more realistic, but using these idealizations makes the mathematics easier.
18 Similar arguments apply for systems in quantum mechanics (Jaffe and Taylor 2018, 405).
19 There is an analogous argument for fields. The mathematics is more advanced, but the overall logic of conditionality is the same (Pitts 2019, 695–99). While this Lagrangian test fails in general relativity (Pitts, private correspondence), the question can be reframed in terms of Killing fields, as we discussed in the previous subsection.
20 While Papineau instead uses an inductive argument in the passage cited, "in principle" language creeps in elsewhere (Papineau 2009, 54).
21 The spontaneous collapse of a quantum wave packet predicted in the GRW interpretation would not conserve momentum or energy (Gao 2017, 145; Maudlin, Okon, and Sudarsky 2020, 74–75).
22 According to Papineau critic E. J. Lowe, physicalists appeal to conservation less often these days since it does not close the argument the way they want (2010, 61).
23 See Smith (2000) and Price and Corry (2006) for more on the argument that physics replaced causation with laws of nature.
24 This is related to Carroll's coupling problem (Section 1.3.3). Pitts answers it in part in (2020, 1938–43).

Works Cited

Bishop, Robert C. 2010. "The Via Negativa: Not the Way to Physicalism." *Mind and Matter* 8 (2): 203–14.

Bishop, Robert C. 2011. "Free Will and the Causal Closure of Physics." In *Visions of Discovery: New Light on Physics, Cosmology, and Consciousness*, edited by Raymond Chiao, 601–11. Cambridge: Cambridge University Press.

Bishop, Robert C., Michael Silberstein, and Mark Pexton. 2022. *Emergence in Context: A Treatise in Twenty-First Century Natural Philosophy*. Oxford: Oxford University Press.

Bunge, Mario. 1980. *The Mind-Body Problem: A Psychobiological Approach*. Foundations and Philosophy of Science and Technology Series. Oxford: Pergamon Press.

Butterfield, Jeremy. 1998. "Quantum Curiosities of Psychophysics." In *Consciousness and Human Identity*, edited by John Cornwell, 122–59. Oxford: Oxford University Press.

Carroll, Sean. 2010. "Energy Is Not Conserved." *Preposterous Universe* (blog). February 22, 2010. www.preposterousuniverse.com/blog/2010/02/22/energy-is-not-conserved/.

Carroll, Sean. 2011. "Physics and the Immortality of the Soul." Scientific American Blog Network. May 23, 2011. https://blogs.scientificamerican.com/guest-blog/physics-and-the-immortality-of-the-soul/.

Carroll, Sean. 2016a. "Consciousness and Downward Causation." *Preposterous Universe* (blog). September 8, 2016. www.preposterousuniverse.com/blog/2016/09/08/consciousness-and-downward-causation/.

Carroll, Sean. 2016b. *The Big Picture: On the Origins of Life, Meaning, and the Universe Itself*. New York: Dutton.

Carroll, Sean. 2019. *Spacetime and Geometry: An Introduction to General Relativity*. Cambridge: Cambridge University Press.

Carroll, Sean, and Jackie Lodman. 2021. "Energy Non-Conservation in Quantum Mechanics." *Foundations of Physics* 51 (4): 1–15. https://doi.org/10.1007/s10701-021-00490-5.

Cartwright, Nancy. 2011. "Natural Laws and the Closure of Physics." In *Visions of Discovery: New Light on Physics, Cosmology, and Consciousness*, edited by Raymond Y. Chiao, Marvin L. Cohen, Anthony J. Leggett, William D. Phillips, and Charles L. Harper, 612–22. Cambridge: Cambridge University Press.

Collins, Robin. 2008. "Modern Physics and the Energy-Conservation Objection to Mind-Body Dualism." *American Philosophical Quarterly* 45 (1): 31–42.

Collins, Robin. 2011. "The Energy of the Soul." In *The Soul Hypothesis: Investigations into the Existence of the Soul*, edited by Mark C. Baker and Stewart Goetz, 123–33. New York: Continuum.

Crowell, Benjamin. 2022. *General Relativity*. Fullerton: Fullerton College. https://phys.libretexts.org/@go/page/11455.

Cucu, Alin. 2022. "Interacting Minds in the Physical World." Doctoral Dissertation, University of Lausanne, Lausanne, Switzerland.

Cucu, Alin, and J. Brian Pitts. 2019. "How Dualists Should (Not) Respond to the Objection from Energy Conservation." *Mind and Matter* 17 (July): 95–121.

Dennett, Daniel. 1991. *Consciousness Explained*. Boston: Little, Brown and Co.

Fitzpatrick, Simon. 2013. "Simplicity in the Philosophy of Science." In *Internet Encyclopedia of Philosophy*, edited by James Fieser and Bradley Dowden. https://iep.utm.edu/simplici/.

Gao, S. 2017. *The Meaning of the Wave Function: In Search of the Ontology of Quantum Mechanics*. Cambridge: Cambridge University Press.

Gillett, Carl, and Barry Loewer, eds. 2001. *Physicalism and Its Discontents*. Cambridge: Cambridge University Press.

Harrison, Edward Robert. 2000. *Cosmology: The Science of the Universe*, 2nd ed. Cambridge: Cambridge University Press.

Hossenfelder, Sabine. 2018. "10 Physics Facts You Should Have Learned in School but Probably Didn't." *BackReAction* (blog). July 30, 2018. http://backreaction.blogspot.com/2018/07/10-physics-facts-you-should-have.html.

Hossenfelder, Sabine. 2019. "The Case for Strong Emergence." In *What Is Fundamental?*, edited by Anthony Aguirre, Brendan Foster, and Zeeya Merali, 85–94. The Frontiers Collection. Cham: Springer International Publishing. https://doi.org/10.1007/978-3-030-11301-8_9.

Hossenfelder, Sabine. 2020. "You Don't Have Free Will, but Don't Worry." *BackReAction* (blog). October 10, 2020. http://backreaction.blogspot.com/2020/10/you-dont-have-free-will-but-dont-worry.html.

Jaffe, Robert L., and Washington Taylor. 2018. *The Physics of Energy*. Cambridge: Cambridge University Press.

José, Jorge Valenzuela, and Eugene Jerome Saletan. 2006. *Classical Dynamics: A Contemporary Approach*. Cambridge: Cambridge University Press.

Kim, Jaegwon. 1996. *Philosophy of Mind*. Dimensions of Philosophy Series. Boulder: Westview Press.

Kim, Jaegwon. 1998. *Mind in a Physical World: An Essay on the Mind-Body Problem and Mental Causation*. Representation and Mind. Cambridge: MIT Press.

Lowe, E. J. 2003. "Physical Causal Closure and the Invisibility of Mental Causation." In *Physicalism and Mental Causation*, edited by Sven Walter and Heinz-Dieter Heckmann, 137–54. Charlottesville: Imprint Academic.

Lowe, E. J. 2010. *Personal Agency: The Metaphysics of Mind and Action*. Oxford: Oxford University Press.

Maudlin, Tim, Elias Okon, and Daniel Sudarsky. 2020. "On the Status of Conservation Laws in Physics: Implications for Semiclassical Gravity." *Studies in History and Philosophy of Science Part B: Studies in History and Philosophy of Modern Physics* 69 (February): 67–81. https://doi.org/10.1016/j.shpsb.2019.10.004.

Mohrhoff, Ulrich. 1997. "Interactionism, Energy Conservation, and the Violation of Physical Laws." *Physics Essays* 10 (4): 651–65.

Montero, Barbara. 2003. "Varieties of Causal Closure." In *Physicalism and Mental Causation*, edited by Sven Walter and Heinz-Dieter Heckmann, 173–87. Charlottesville: Imprint Academic.

Noether, Emmy. (1918) 1971. "Invariant Variation Problems." Translated by M. A. Tavel. *Transport Theory and Statistical Physics* 1 (3): 186–207. https://doi.org/10.1080/00411457108231446.

Papineau, David. 2000. "The Rise of Physicalism." In *The Proper Ambition of Science*, edited by M. W. F. Stone and Jonathan Wolff, 174–208. London Studies in the History of Philosophy, Vol. 2. New York: Routledge.

Papineau, David. 2002. *Thinking about Consciousness*. Oxford: Clarendon Press.

Papineau, David. 2008. "Must a Physicalist Be a Microphysicalist?" In *Being Reduced*, edited by Jakob Hohwy and Jesper Kallestrup. Oxford: Oxford University Press. https://doi.org/10.1093/acprof:oso/9780199211531.001.0001.

Papineau, David. 2009. "The Causal Closure of the Physical and Naturalism." In *The Oxford Handbook of Philosophy of Mind*, edited by B. McLaughlin, A. Beckermann, and S. Walter, 53–65. New York: Oxford University Press.

Papineau, David. 2015. "There Is No Trace of Any Soul Linked to the Body." In *The Myth of an Afterlife: The Case against Life after Death*, edited by Michael Martin and Keith Augustine, 369–76. Lanham: Rowman & Littlefield.

Pitts, J. Brian. 2019. "Conservation Laws and the Philosophy of Mind: Opening the Black Box, Finding a Mirror." *Philosophia* 48: 673–707. https://doi.org/10.1007/s11406-019-00102-7.

Pitts, J. Brian. 2020. "General Relativity, Mental Causation, and Energy Conservation." *Erkenntnis* 26 (3): 1931–73. https://doi.org/10.1007/s10670-020-00284-7.

Pitts, J. Brian. 2021. "Conservation of Energy: Missing Features in Its Nature and Justification and Why They Matter." *Foundations of Science* 26 (3): 559–84. https://doi.org/10.1007/s10699-020-09657-1.

Price, Huw, and Richard Corry. 2006. "A Case for Causal Republicanism?" In *Causation, Physics, and the Constitution of Reality: Russell's Republic Revisited*, edited by Huw Price and Richard Corry, 1–10. Oxford: Oxford University Press.

Rickabaugh, Brandon, and James Porter Moreland. 2023. *The Substance of Consciousness: A Comprehensive Defense of Contemporary Substance Dualism*. Chichester: Wiley-Blackwell.

Rivat, Sébastien. 2024. "Open Systems across Scales." *Synthese* 205 (1): 11. https://doi.org/10.1007/s11229-024-04817-3.

Robb, David, and John Heil. 2021. "Mental Causation." In *Stanford Encyclopedia of Philosophy*, edited by Edward N. Zalta, Spring 2021. Metaphysics Research Lab, Stanford University. https://plato.stanford.edu/archives/spr2021/entries/mental-causation/.

Robinson, William. 2023. "Epiphenomenalism." In *Stanford Encyclopedia of Philosophy*, edited by Edward N. Zalta and Uri Nodelman, Summer 2023. Metaphysics Research Lab, Stanford: Stanford University. https://plato.stanford.edu/archives/sum2023/entries/epiphenomenalism/.

Russell, Bertrand. 1913. "On the Notion of Cause." *Proceedings of the Aristotelian Society* 13 (1): 1–26. https://doi.org/10.1093/aristotelian/13.1.1.

Sapolsky, Robert M. 2023. *Determined: A Science of Life without Free Will*. New York: Penguin Press.

Searle, John R. 2004. *Mind: A Brief Introduction*. Fundamentals of Philosophy Series. Oxford: Oxford University Press.

Smith, Sheldon R. 2000. "Resolving Russell's Anti-Realism about Causation." *The Monist* 83 (2): 274–95.

Wilson, Jessica M. 2021. *Metaphysical Emergence*. Oxford: Oxford University Press.

6 Reductions

6.1 Two Approaches

If "reduction" isn't the most familiar term, the idea certainly is. We know that pressure—something we can sense—is caused by the force exerted by gas molecules colliding with the walls of a container. We know that sound is carried by compression waves traveling through the air. We know the molecular composition of millions of organic substances and how genes are responsible for many biological traits. And we know that all the varied types of matter we experience are made up of the elements on the periodic table. Each of these can be described as a reduction, as in "pressure has been reduced to molecular collisions in a gas."

This is, of course, a short list. The tally of successful reductions is impressive. Reductions help us to understand the world better, seeing how complexity and diversity arise from relative simplicity. One might even think that reduction is the central methodological principle of science—perhaps the very thing that distinguishes it from medieval natural philosophy. It might even seem inevitable that science will one day reduce everything to particle physics. That's certainly the picture most students of the various sciences come away with.

How well this picture maps onto reality depends in part on precisely what "reduction" means. As we will see, it is an ambiguous term. Moreover, many philosophers and scientists now believe that reduction has limits. These limits have been expressed in an alternative approach known as "emergence," an equally ambiguous term. The goal of this chapter and the next is to get a handle on all this. Two chapters can only provide an overview of this unruly terrain, but that will be enough for our purposes. Given the diversity of views in this chapter, the reader may wonder what these approaches have in common. How can they all be called "reductions." While I sympathize, the goal here is mostly descriptive. I will not attempt to impose order from on high and take no stand

DOI: 10.4324/9781003681199-6

on what the exact components of reduction are: concepts, say, or theories, models, explanations, entities, processes, systems, etc.

In terms of physics and free will, the scientists and philosophers discussed in Chapter 1 are reductionists of some type or other. They believe that consciousness in general and our sense of free will in particular can be reduced to brain activity. This chapter considers the different ways one might understand such claims.

6.2 Philosophy and Science on Reduction

Philosophers and scientists have approached reduction in their own disciplinary ways. Analytic philosophers start with those things used to represent natural systems, such as theories, models, and concepts. Reduction for them is usually about connecting and even deriving higher-level theories (or models or concepts) from lower-level ones, where different "levels" more or less correspond to a hierarchy within the natural sciences. We will call this approach *epistemological reduction*. There are several varieties. Mapping them will take up most of this chapter.

Scientists, on the other hand, often focus on entities, properties, and events themselves, not their representations (van Gulick 2001, 3–4). Their approach is more about how things are, regardless of whether or not we know how to derive one set of statements from another. We will call this approach *ontological reduction*. A naïve (but common) version says that X is nothing but Y. For example, a brain is nothing but a collection of neurons, a liquid is nothing but a collection of molecules, and ultimately, any given physical object is nothing but the particles in the Standard Model. We saw some examples of this in Chapter 1:

> A lot of people seem to think it is merely a philosophical stance that the behavior of a composite object (for example, you) is determined by the behavior of its constituents—that is, subatomic particles... But reductionism—according to which the behavior of an object can be deduced from ("reduced to," as the philosophers would say) the properties, behavior, and interactions of the object's constituents—is not a philosophy. It's one of the best-established facts about nature.
>
> (Hossenfelder 2022, 82–83)

Carroll puts it this way:

> You and all your friends and loved ones, not to mention your casual acquaintances and sworn enemies, are combinations of ... matter particles interacting with each other and ... forces according to the rules of quantum field theory. Every book you've ever read, every

> meal you've ever eaten, every painting you've seen and song you've heard have arisen out of the collective dynamics of these basic pieces... There is still an enormous amount of work to be done to show how these microscopic constituents come together to produce the wonders of the macroscopic world, but the list of ingredients is well understood.
>
> (Carroll 2024, 265)

Hossenfelder addresses the same theme again elsewhere:

> [Laws] of nature apply to you and your brain because you are made of particles, and what happens with you is a consequence of what happens with those particles. A lot of people seem to think this is a philosophical position. They call it "materialism" or "reductionism" and think that giving it a name that ends on –ism is an excuse to not believe it. Well, of course you can insist to just not believe reductionism is correct. But this is denying scientific evidence. We do not guess, we know that brains are made of particles.
>
> (Hossenfelder 2020b)

I have been using "physicalism" to mean roughly the same thing as "materialism." Hossenfelder conflates these with reductionism.[1] The confusion is understandable. Ontological reductionism relies to some extent on physicalism. Alexander Rosenberg makes the point here about cell division, but it generalizes: "Reductionists will appeal to physicalism—the thesis that the physical facts fix all of the facts—to underwrite the thesis that every [process within cell division] is a physical one" (Rosenberg 2020, 26). Methodologically, this means that science is continually drawn to smaller scales, as Steven Weinberg explains:

> Now reductionism, as I've described it in terms of the convergence of arrows of explanation, is not a fact about scientific programs, but a fact about nature. I suppose if I had to give a name for it, I could call it objective reductionism. It is very far from a truism. In particular, these arrows of explanation might have led to many different sources... This discovery, that the arrows of explanation point down to a common source, is quite new.
>
> (Weinberg 2001, 18)

Weinberg makes it clear that he is talking about nature itself, not merely the relation between scientific theories. If physical objects are ultimately all that exist, then somehow or other all higher-level entities must be reducible to lower ones.

What then of the "special sciences"—those other than physics, such as chemistry and biology? Physicalists accept them insofar as they offer convenient ways of referring to configurations of more basic entities. Chemists will sometimes say that biology is really just applied chemistry. Physicists will sometimes say that chemistry is really just applied physics. Those are more aspirational slogans than analysis, but the idea is clear enough.

More precisely, one might say that higher-level properties (or entities, events, systems, etc.) depend on and are determined by the lower-level properties of which they are composed. Take any molecule, like carbon dioxide CO_2. Ontological reduction says that the properties of that molecule and its gaseous collections are completely determined by its constituent atoms in a given configuration. In general, "Properties and behaviors in a particular (lower 'level' or smaller scale) domain ... offer all conditions sufficient to determine properties and behaviors in another domain" (Bishop, Silberstein, and Pexton 2022, 27).

It might not initially sound like there is much difference between epistemic and ontological reduction. While there is quite a bit of overlap, the two can come apart. This will become evident in Section 6.4 when we discuss the limits of epistemic reduction. Ontological reduction is the simpler of the two. Whatever entities inhabit the ground floor of reality—subatomic particles, quantum fields, strings, etc.—ontological reduction says that all higher-level stuff is somehow made up of those entities. That leaves a lot of ambiguity when it comes to epistemic reduction. We will consider three types in the following sections before turning to problems confronting each.

6.3 Epistemic Reduction

6.3.1 Theory Reduction

Much of the philosophical work on reduction traces back to the twentieth century and logical empiricism, with its (excessive) focus on language. The logical empiricists' question was whether the sentences composing a given scientific theory could be translated into those composing a lower-level theory. Finding such translations was considered a mark of scientific progress (Hüttemann and Love 2016, 462). Once complete, the reducing theory would explain why the reduced theory has the form that it does. Molecular physics, for example, not only shows that air is composed of molecules, but also explains thermodynamic temperature and pressure. Many hoped that the sciences would one day be fully unified, with all such translations made evident (Oppenheim and Putnam 1958).

Philosopher Ernest Nagel made significant advances on this project (Nagel 1961; Bickle 2019, 66–70). Theories for Nagel are expressed by statements arranged in the hierarchy of scientific disciplines—a "layer-cake" view.[2] A theory in one discipline is successfully reduced when one shows how to derive its experimental laws from a theory in another discipline lower in the hierarchy. The classic example is the reduction of thermodynamics to statistical mechanics. With the derivation in hand, one can see how the more fundamental laws explain those in the reduced theory.

The main obstacle for this approach is that the vocabularies in the reduced and reducing theories seldom match. For example, temperature is a theoretical term in thermodynamics that does not appear in statistical mechanics. This makes a straightforward logical derivation impossible (Nagel 1961, 352). What was needed were so-called "bridge laws" connecting the predicates in one theory to those in another. The reduction of the theory of optics to electromagnetic radiation, for example, requires a bridge law to the effect that light waves just are a type of electromagnetic wave. This allows statements about light waves to be expressed in terms of the reducing theory so that the derivation can proceed. In general, however, it was easier to talk about bridge laws abstractly than to find them, or even to say precisely what they might be (Batterman 2002, 63). Are bridge laws discovered, like other laws of nature? Do we instead invent them as a matter of convention? Opinions varied.

While Nagel's view has seen many amendments, philosophers today are less enthusiastic about theory reduction, although some still favor the basic schema (van Riel and Van Gulick 2024, sec. 2.2).

6.3.2 Limit-Case Theory Reduction

Physicists sometimes use "reduction" in a different way, what Thomas Nickles calls reduction$_2$ (1973). Reduction$_2$ relates theories at the same level, showing how a newer theory can take the form of an older one under the right circumstances.[3] For example, one might say that special relativity reduces to Newtonian mechanics at small velocities. Our experience is well-described by Newtonian mechanics because the unique contribution of relativity cannot be detected at everyday speeds. The more velocities approach zero, the smaller the observable difference between the two theories. Mathematically, this can be expressed as the limit where $v \rightarrow 0$. One might likewise say that quantum mechanics reduces to classical mechanics as $\hbar \rightarrow 0$, where $\hbar$ is Planck's constant divided by 2π. This is one way to express the correspondence principle, such that large quantum systems behave like classical ones. Quantum effects are essentially undetectable in macroscopic objects. As $\hbar$ in Schrödinger's equation approaches zero, the wave function's behavior resembles that of a classical particle.[4]

Note that it is the successor theory in limit-case reductions that reduces to the older theory (Palacios 2022, sec. 2.5). This reverses the direction of theory reductions in the previous section, where it was the older, macroscopic theory that was reduced to a newer, underlying theory. The goal of limit-case reduction is not to explain a higher-level theory, but to show why a previous theory continues to be so useful, given that the successor theory had been true all along.

Same-level limit reductions capture one way physicists talk about reduction and so deserve to be mentioned. But it applies to such a small number of cases that it cannot tell us much about reduction overall.

Batterman shows that limit reductions require a particular mathematical relation between theories and that the subtle failure of this relation can itself be significant in physics (Batterman 1995, sec. 3; Batterman 2002, 19, 122–23). In so-called "singular limits," the behavior of the system *in* the limit $(a = 0)$ is qualitatively different from that as the limit is approached $(a \to 0)$. Let's start with an analogy by physicist Michael Berry (2002b). If you bite into an apple and find ¾ of a remaining worm, that's bad. If you had found ½ of a worm, that would be worse. If you found ¼ of a worm, that would be worse still. But if you found no parts of a worm whatsoever, that would be a good thing. Reaching the limit, zero, has a very different result than the approach to the limit. This is relevant to the supposed reduction of thermodynamics. Statistical mechanics predicts phase transitions, such as gas condensation, but only in systems with an infinite number of particles.[5] It does not allow for phase transitions in large but finite systems. The relevant equations predict one behavior in the *approach* to the thermodynamic limit. *In* the limit of infinite particles—which is obviously an idealization—something quite different happens. This is the puzzling nature of singular limits.

For our purposes, the discontinuity between approaching and being in the limit again shows that the reduction of thermodynamics to statistical mechanics is a more complicated matter than advertised. If the reduction doesn't work in finite systems, and all systems are finite, that would leave a significant gap between the two levels.

6.3.3 Explanatory Reduction

As the name suggests, explanatory reductions explain upper-level behavior/properties in terms of underlying causes, processes, etc. (Hüttemann and Love 2016, sec. 4.2). These reductions are most prominent in the biological sciences, where the idea of *mechanism* is often used (Bickle 2006; Sarkar 1998). Cell division, for example, is explained in terms of both identifiable stages and the molecular mechanisms involved, such as the formation of chromosomes. The activation of neurons is explained as channels of ion movements (sodium and potassium) and electrical signal propagating (Machamer, Darden, and Craver 2000, 8–13). Such reductions proceed piecemeal, one set of mechanisms at a time. There is no implicit expectation that whole theories will one day be reduced. This in turn makes for fewer "in principle" qualifiers in biology (Brigandt and Love 2023, sec. 3.2), such as "An electron or a quark may not have a conductivity, but *in principle* you can calculate how they form atoms, and molecules, and metals, and then the conductivity is a consequence of this" (Hossenfelder 2020a) Italics added. As Batterman notes, while reductionists are quick to make such claims, they are "rarely, if ever, accompanied by a suggestion as to how this might be done" (2021, 10).

One controversy in the philosophical literature is whether the discovery of mechanisms should count as reduction. Some say no. Explaining an engine in terms of its parts isn't reducing the engine. If there is no sense in which higher-level properties are derivable from lower ones, that's not reduction. Others believe that insofar as one is understanding higher-level phenomena in terms of smaller scales, then the appeal to mechanisms is consistent with reduction. The explanatory arrows point down, in Weinberg's sense.

Before moving on, we should note an extreme form of epistemic reduction known as *eliminativism*. If one theory (or even subdiscipline) can be fully reduced, perhaps there will no longer be a need for the reduced theory. Successful reduction means the theory will fade away, like the caloric theory of heat.[6] Philosophers Paul and Paula Churchland argued that folk psychology, using mental terms like "belief" and "desire," will be eliminated once neuroscience is sufficiently advanced (Churchland 1988). Eliminativism has not aged well. Not only does it seem naïve, it does not account for the limits to epistemic reduction to be discussed in the next section.

Having examined these three types of epistemic reduction, we can now see how they differ in their approach to scientific explanation. Theory reduction focuses on the logical derivation of higher-level theories from lower-level ones, requiring bridge laws to connect disparate vocabularies across scientific disciplines. Limit-case reduction instead shows how newer theories approximate older ones under specific conditions, explaining why older theories remain useful despite their limitations. Explanatory reduction takes a more mechanistic approach, explaining higher-level phenomena through their constituent parts and processes without necessarily attempting complete theoretical derivation. These approaches each capture something important about scientific practice, but none provides a complete picture of how reduction operates across the sciences. The logical empiricists' focus on theory reduction emphasizes derivability but struggles with bridge laws, limit-case reduction helps us understand successive theories but applies only where mathematical limits can be defined, and explanatory reduction aligns with scientific practice in fields like biology but proceeds piecemeal rather than offering comprehensive theoretical connections. As we will see in the next section, each type of reduction faces significant limitations that challenge the reductionist program.

This is a good place to reintroduce a term from the preface: fundamentalism. The fundamentalist mindset draws on several themes found in this chapter. First, there is a hierarchy of theories, with some being more "fundamental" than others. The most fundamental theory is at the lowest level and serves as a "ground" for all higher-level theories. Second, all phenomena at higher levels can, in principle, be perfectly described, predicted, and explained by the fundamental, structural theories. Any obstacles to this are seen as epistemic limitations, not as a failure of the fundamental theory. Third, the most fundamental theory is also the most epistemically trustworthy, providing the most certain and verifiable claims about the world. These three show why fundamentalist physicists don't consult with the other sciences on questions about reality, time, or free will. Those sciences don't have much to add to the conversation. Fundamentalism thus represents a kind of physicalism that combines ontological reduction (all phenomena reduce to fundamental physics), explanatory reduction (all explanation must be bottom-up), and epistemic privilege (fundamental physics provides the most reliable knowledge).

6.4 Limitations and Failures

None of the three types of epistemic reduction have a claim to being the real or correct view of reduction. I doubt most scientists have an unambiguous concept of what "reduction" means. They just think of it as a good approach that often leads to the explanations they want. It is a general, methodological idea with a track record of success.

We move on to the part of the story that is generally left out. Failures to reduce are often seen as temporary setbacks, with the history of science purportedly showing that, in due time, every higher-level entity will be reduced to those at lower levels. Matters are not so simple. The issues raised in this section are not aimed at any one type of reduction but rather the reductive impulse—the sense that reduction is simply the way science advances. As we will see, the grand reductive narrative relies on selective historical examples and incomplete accounts of scientific practice. Once the details are filled in, that "track record of success" is less compelling.

6.4.1 Oversimplified Examples

Examples from history are typically used to make a point. Narrators, including science educators, often leave out the parts of the story that muddy the road to their conclusion. Consider the supposed reduction of thermodynamics to statistical mechanics. One oft-repeated claim is that temperature is really just the average kinetic energy of molecules. At best, this is true for ideal gases and straightforwardly only applies to gases at equilibrium. The temperature of a candle is an entirely different matter. Moreover, the derivation only works at the "thermodynamic limit," where the number of particles and volume approach infinity while the density is held constant. And while nonequilibrium systems evolve from low to high entropy states, there is still a debate about how this directionality should be understood in terms of molecular interactions, which have no preferred direction.

Limitations like this apply to most of the standard examples of the reduction of thermodynamics (Sklar 2009). Its full reduction is something of a myth. The promise that such reductions could still be carried out "in principle" cannot bear the burden:

> For the moment, however, let us merely note a characteristic fallacy that often lies behind these glib "in principle" assurances… [Philosophers] often appeal to the notable successes of what Mazenko calls "traditional statistical mechanics," viz. standard approaches to ideal gases and perfect crystal arrays. But single victories do not insure the success of an extended campaign. We shouldn't generalize from the real but limited successes of "traditional statistical mechanics" to the conclusion that the same pristine methods will work "in principle" everywhere. That supposition overlooks the fact that these bottom-up success stories rely heavily upon the absence of higher-scale irregularities within a gas or perfect crystal.
>
> (Wilson 2017, 231–32)

Classical statistical mechanics idealizes away intermediate (meso) level structure that will become crucial in Chapter 8. In a sense, that's fine. Idealizations often help us get a handle on the big picture. In this case, it is misleading. One cannot generalize from the success of statistical mechanics and so is a bad model to follow.

Consider a similar example. "You can derive the laws of hydrodynamics (which describe the motions of fluids) from the theory of atoms" (Hossenfelder 2022, 86). Not in general, you can't. Doing so, again, typically requires that the fluid be in equilibrium. Furthermore, there are two idealizations that Hossenfelder fails to mention. First, the matter in the fluid is considered to be a true continuum rather than atomic—more like molasses than sand.[7] Second, the fluid has no viscosity. These idealizations do not merely make the derivation that Hossenfelder claims easier; they make it possible. No one has ever derived the whole of hydrodynamics from atomic physics using realistic assumptions. In fact, once a system gets larger than ten particles, physicists can only make approximations about higher level properties, which "require experimental input and specific, local details. What this indicates is a breakdown of the reductionist ideal of deriving explanations of a large number of phenomena from a few simple equations or laws" (Morrison 2015, 97). For anyone who thinks otherwise, the Clay Mathematics Institute offers a $1 million prize for a general solution to the Navier-Stokes equations in three dimensions (www.claymath.org/millennium-problems/).[8] Good luck! Carroll concedes that a "microscopic theory can be said to entail the macroscopic one; but that's often something we take for granted, not something that can explicitly be demonstrated" (2016, 100).

6.4.2 *Context Dependence*

How and when a higher-level property manifests itself often depends on more than its material components. It also depends on its environment and history. Consider hysteresis, a common problem in engineering. A newly manufactured metal bar has a given *flexural strength*, defined as the highest stress the bar can undergo without bending. Now say the bar is bent to a 90-degree angle and then restored to its original shape. The strength of the bar is now measurably lower. Flexural strength depends on the past, not merely on molecular composition. Hysteresis also describes the way your smartphone battery degrades over time. The phenomenon is not something that can be modeled in a bottom-up fashion (Wilson 2017, 221).

Philosophers of biology Ingo Brigandt and Alan Love give another example:

> One key challenge for reduction stems from the fact that the effect of a molecular entity or mechanism may depend strongly on the context in which it occurs. Classical geneticists were fully aware of the fact that a phenotype is brought about by the interaction of several (many) classical genes—*the same allele may lead to two different phenotypes if occurring in two individuals with a different overall genotype.*
>
> (Brigandt and Love 2023, sec. 4.1, italics added, citations removed)

Instead of genes alone determining observable characteristics, the overall genetic makeup of an individual determines how some genes manifest themselves. This challenges the reductive story that properties at the level of organism (phenotype) can be reduced to the organism's genetic makeup (genotype). This is not news to biologists. They understand the complexities of living systems at all levels. The problem is pedagogical. With the limited time science educators have with most students, they focus on what they take to be the main narrative and leave the complications for more advanced study.

Context dependence will be a key issue for Bishop et al. in Section 7.4 (2022). Here it shows that higher-level properties cannot be wholly reduced to lower-level properties present at a given time. Higher-level properties often depend on their environment and history.

6.4.3 Multiple Realization

There is often no single base property to which higher-level properties can be reduced. Consider Hilary Putnam's peg and board example (1975, 295–96). Putnam asks us to imagine a wooden board with two holes in it: a circular hole with a 1-inch diameter and a square hole that is 1 inch on each side. Now, imagine two pegs: a circular one with a diameter slightly less than 1 inch and a square one with sides slightly less than 1 inch. The circular peg will pass through the circular hole, but the square peg will not.

Putnam then considers two possible explanations for this phenomenon. The first is geometric. The dimensions of the holes and the pegs explain why the square peg won't fit. The second is compositional. The properties of the constituent atoms of the wood prevent the square peg from passing through. Putnam then argues that the compositional explanation is largely irrelevant. Why? Because the board could have been made of a completely different material. The composition of the wood does not explain the behavior of the pegs. The correct, more general explanation lies at a higher level.

This illustrates a general point. Higher-level traits can result from many different micro-realizers (Bechtel and Richardson 2010, xviii–xix, xxx; Polger and Shapiro 2016; Brigandt and Love 2023, sec. 4.2). For example, the ability to see is instantiated in many ways among animals, including camera-style eyes in mammals, compound eyes in insects, and mirrored eyes in scallops. A given protein like hemoglobin can be realized by different amino acid sequences. Logic gates (AND, OR, NAND, etc.) can be constructed using relays, vacuum tubes, or transistors.

There is no single reductive explanation for multiply realized properties. Even if one has a good explanation for why a given base produces some macro-property, that would not explain why other heterogeneous bases produce the same property. The way in which transistors form a NAND gate does not explain why relays and vacuum tubes can do so as well. What we often want instead is an explanation of the pattern: why these different bases all produce that property. The answer to that question cannot be found in the microconstituents. The phenomenon of multiple realization, therefore, suggests that explanatory power often resides at higher levels, resisting reduction to specific physical substrates.

6.4.4 *Failed Unity*

Going back to the logical empiricists once again, reductionism was often associated with the "unity of science" (Oppenheim and Putnam 1958). The idea was that the gaps between the special sciences would one day be filled in, and we would see how science fits together in a unified hierarchy. Each successful reduction was understood as a closing of one such gap.

The recent history of science, however, remains stubbornly uncooperative. As science has advanced, so has its fragmentation. New specializations (with their attendant journals) arise year by year. Some philosophers of science now endorse a pluralistic, disunified picture for science as a whole (Cartwright 1999, 2000; Kellert, Longino, and Waters 2006).[9] Berry concurs:

> Our understanding of the world is a patchwork of vast scope; it covers the intricate chemistry of life, the sociology of animal communities, the gigantic wheeling galaxies, and the dances of elusive elementary particles. But it is a patchwork nevertheless, and the different areas do not fit well together.
>
> (Berry 2002a, 41)

This patchwork is the result of a failure to reduce. "The most prominent argument against reductionism stems from the observation that straightforward reductions hardly ever occur. Hence, reductionism cannot be regarded as yielding a coherent picture of what actually goes on in science" (van Riel and Van Gulick 2024, sec. 6).

One indication that philosophers were having doubts about reductionism and the unity of science was the widespread use of the term "supervenience" in the 1970s and 1980s. If some property (or fact, event, etc.) supervenes on another, that signals a kind of dependence. There can be no difference at the supervenient level without some difference among the base properties. Philosopher Donald Davidson's classic example was that mental states supervene on physical states (1980, 214). In order for a person's mental state to change, there would have to be a change in that person's body, especially the brain. However, supervenience fails to specify the nature of that dependence. If successful reductions had been forthcoming, there would have been no need to tolerate this vagueness. The relation between levels would have been made explicit. (Supervenience has more recently fallen out of favor.)

Ironically, philosophers have tended to think of the relation between levels in terms of epistemic reduction, especially theory reduction. Nonetheless, most no longer believe that full-blown theory reduction is possible. Today, many philosophers consider themselves "non-reductive physicalists."[10] "[The] reigning view in mainstream philosophy is consequently that reductionism is basically a *dead*, and perhaps even somewhat *distasteful*, position" (Gillett 2016, 4). Hüttemann and Love highlight the failure of Nagelian reductions in particular as driving this change (2016, 465).

Having examined these four limitations to reduction, we can now see why the grand reductionist narrative is more complicated than it first appears. The oversimplified examples commonly used to illustrate reduction often hide significant idealizations and contextual constraints that limit their applicability. Context dependence reveals that higher-level properties frequently depend not just on their material components but also on their environment and history. Multiple realization demonstrates that the same higher-level property can be instantiated by heterogeneous lower-level bases, making a single reductive explanation impossible. And the failed unity of science—contrary to the expectations of logical empiricists—shows that as science has advanced, it has become more fragmented rather than more unified.

Taken together, these limitations suggest that reduction cannot serve as the sole methodological principle for scientific progress. The patchwork nature of our scientific understanding, with its various specialized domains that don't always fit neatly together, indicates that we need alternative approaches to reduction. With that, we now turn to what is currently the strongest rival to reductionism: emergence.

Notes

1 Materialism and physicalism are ontological claims. Reduction in its many forms tends to be relational, e.g., the relation between theories at different levels.

2 The layer-cake analogy is more often associated with Oppenheim and Putman (1958), although they did not explicitly use that language.

3 Saying that one theory is being reduced to another is a bit imprecise. The relation is between equations within the two theories. This is quite different from the inter-level reductions in the previous section (Hüttemann and Love 2016, 468).

4 This example is a bit more nuanced than the first since a physical constant cannot "go to zero." Instead, the goal is to see how quantum mechanical equations behave if the quantization effects become infinitesimally small—one common way physicists think of the correspondence principle.

5 A system with a finite number of particles, no matter how large, can never truly exhibit the sharp mathematical discontinuities found in phase transitions.

6 The idea was that heat was due to a fluid called "caloric." While successful for a time, even as a basis for Sadi Carnot's work in thermodynamics, it was replaced by a mechanical understanding of heat.

7 This analogy for continuum and particle systems is sometimes attributed to Bertrand Russell.
8 More precisely, they are not asking for a solution but a general proof of the existence and uniqueness of solutions (Earman 2007, 1384).
9 Shapiro argues for a way to rehabilitate the idea of unity of science, but not in a way that boosts the prospects of epistemic reduction (Shapiro 2022).
10 According to O'Connor and Wong, non-reductive physicalism has split into two groups (2005, 661–62). The one described here has the same metaphysic as reductive physicalism. The other group affirms the reality of higher-level properties, something beyond the level of physics. They take the latter group to hold a kind of soft property dualism (Robinson 2023, sec. 2.2). I think this latter group is better characterized as emergentists.

Works Cited

Batterman, Robert W. 1995. "Theories between Theories: Asymptotic Limiting Intertheoretic Relations." *Synthese* 103 (2): 171–201. https://doi.org/10.1007/BF01090047.

Batterman, Robert W. 2002. *The Devil in the Details: Asymptotic Reasoning in Explanation, Reduction, and Emergence*. Oxford: Oxford University Press.

Batterman, Robert W. 2021. *The Middle Way: A Non-Fundamental Approach to Many-Body Physics*. Oxford: Oxford University Press.

Bechtel, William, and Robert C. Richardson. 2010. *Discovering Complexity: Decomposition and Localization as Strategies in Scientific Research*. Cambridge: MIT Press.

Berry, Michael. 2002a. "Chaos and the Semiclassical Limit of Quantum Mechanics (Is the Moon There When Somebody Looks?)." In *Quantum Mechanics: Scientfic Perpectives on Divine Action*, edited by Robert J. Russell, Philip Clayton, Kirk Wegter-McNelly, and John Polkinghorne, 41–54. College Park: Center for Theology and the Natural Sciences.

Berry, Michael. 2002b. "Singular Limits." *Physics Today* 55 (5): 10–11. https://doi.org/10.1063/1.1485555.

Bickle, John. 2006. "Reducing Mind to Molecular Pathways: Explicating the Reductionism Implicit in Current Cellular and Molecular Neuroscience." *Synthese* 151 (3): 411–34. https://doi.org/10.1007/s11229-006-9015-2.

Bickle, John. 2019. "Reduction." In *The Routledge Handbook of Emergence*, edited by Sophie Gibb, Robin Findlay Hendry, and Tom Lancaster, 65–76. New York: Routledge. https://doi.org/10.4324/9781315675213-5.

Bishop, Robert C., Michael Silberstein, and Mark Pexton. 2022. *Emergence in Context: A Treatise in Twenty-First Century Natural Philosophy*. Oxford: Oxford University Press.

Brigandt, Ingo, and Alan Love. 2023. "Reductionism in Biology." In *Stanford Encyclopedia of Philosophy*, edited by Edward N. Zalta and Uri Nodelman. Stanford: Metaphysics Research Lab, Stanford University. https://plato.stanford.edu/archives/sum2023/entries/reduction-biology/.

Carroll, Sean. 2016. *The Big Picture: On the Origins of Life, Meaning, and the Universe Itself*. New York: Dutton.

Carroll, Sean. 2024. *Quanta and Fields: The Biggest Ideas in the Universe*. New York: Dutton.

Cartwright, Nancy. 1999. *The Dappled World: A Study of the Boundaries of Science*. Cambridge: Cambridge University Press.

Cartwright, Nancy. 2000. "Against the Completability of Science." In *The Proper Ambition of Science*, edited by M. W. F. Stone and Jonathan Wolff, 209–22. London: Routledge.

Churchland, Paul M. 1988. *Matter and Consciousness: A Contemporary Introduction to the Philosophy of Mind*. Rev. ed. Cambridge: MIT Press.

Davidson, Donald. 1980. "Mental Events." In *Essays on Actions and Events*, edited by Donald Davidson. Oxford: Clarendon Press.

Earman, John. 2007. "Aspects of Determinism in Modern Physics." In *Philosophy of Physics*, edited by John Earman and Jeremy Butterfield, 1369–1434. Handbook of the Philosophy of Science. Vol 2. Amsterdam: Elsevier.

Gillett, Carl. 2016. *Reduction and Emergence in Science and Philosophy*. Cambridge: Cambridge University Press. https://doi.org/10.1017/CBO9781139871716.

Hossenfelder, Sabine. 2020a. "What Is Emergence? What Does 'Emergent' Mean?" *Backreaction* (blog). April 3, 2020. http://backreaction.blogspot.com/2020/04/what-is-emergence-what-means-emergent.html.

Hossenfelder, Sabine. 2020b. "You Don't Have Free Will, but Don't Worry." *BackReAction* (blog). October 10, 2020. http://backreaction.blogspot.com/2020/10/you-dont-have-free-will-but-dont-worry.html.

Hossenfelder, Sabine. 2022. *Existential Physics: A Scientist's Guide to Life's Biggest Questions*. New York: Viking.

Hüttemann, Andreas, and Alan Love. 2016. "Reduction." In *The Oxford Handbook in Philosophy of Science*, edited by Paul Humphreys, 460–84. Oxford: Oxford University Press.

Kellert, Stephen H., Helen E. Longino, and C. Kenneth Waters, eds. 2006. *Scientific Pluralism*. Minnesota Studies in the Philosophy of Science. Vol. 19. Minneapolis: University of Minnesota Press.

Machamer, Peter, Lindley Darden, and Carl F. Craver. 2000. "Thinking about Mechanisms." *Philosophy of Science* 67 (1): 1–25.

Morrison, Margaret. 2015. "Why Is More Different?" In *Why More Is Different*, edited by Brigitte Falkenburg and Margaret Morrison, 91–114. The Frontiers Collection. Berlin: Springer. https://doi.org/10.1007/978-3-662-43911-1_6.

Nagel, Ernest. 1961. *The Structure of Science: Problems in the Logic of Scientific Explanation*. New York: Harcourt, Brace & World.

Nickles, Thomas. 1973. "Two Concepts of Intertheoretic Reduction." *The Journal of Philosophy* 70 (7): 181–201. https://doi.org/10.2307/2024906.

O'Connor, Timothy, and Hong Yu Wong. 2005. "The Metaphysics of Emergence." *Noûs* 39 (4): 658–78.

Oppenheim, Paul, and Hilary Putnam. 1958. "Unity of Science as a Working Hypothesis." In *Concepts, Theories, and the Mind-Body Problem*, edited by H. Feigl, M. Scriven, and G. Maxwell, 3–36. Minneapolis: University of Minnesota Press.

Palacios, Patricia. 2022. *Emergence and Reduction in Physics*. Cambridge: Cambridge University Press. https://doi.org/10.1017/9781108901017.

Polger, Thomas W., and Lawrence A. Shapiro. 2016. *The Multiple Realization Book*. Oxford: Oxford University Press. https://doi.org/10.1093/acprof:oso/9780198732891.001.0001.

Putnam, Hilary, ed. 1975. "Philosophy and Our Mental Life." In *Philosophical Papers: Mind, Language and Reality*, Vol. 2, 291–303. Cambridge: Cambridge University Press. https://doi.org/10.1017/CBO9780511625251.016.

Robinson, Howard. 2023. "Dualism." In *The Stanford Encyclopedia of Philosophy*, edited by Edward N. Zalta and Uri Nodelman. Stanford: Metaphysics Research Lab, Stanford University. https://plato.stanford.edu/archives/spr2023/entries/dualism/.

Rosenberg, Alexander. 2020. *Reduction and Mechanism*. Cambridge: Cambridge University Press. https://doi.org/10.1017/9781108592949.

Sarkar, Sahotra. 1998. *Genetics and Reductionism*. Cambridge Studies in Philosophy and Biology. Cambridge: Cambridge University Press.

Shapiro, Lawrence. 2022. "Rethinking the Unity of Science Hypothesis: Levels, Mechanisms, and Realization." In *Levels of Reality in Science and Philosophy*, edited by Stavros Ioannidis, Gal Vishne, Meir Hemmo, and Orly Shenker, 209–27. Jerusalem Studies in Philosophy and History of Science. Cham: Springer International Publishing.

Sklar, Lawrence. 2009. "Philosophy of Statistical Mechanics." In *Stanford Encyclopedia of Philosophy*, edited by Edward N. Zalta. Stanford: Metaphysics Research Lab, Stanford University. http://plato.stanford.edu/entries/statphys-statmech/.

van Gulick, Robert. 2001. "Reduction, Emergence and Other Recent Options on the Mind/Body Problem: A Philosophic Overview." *Journal of Consciousness Studies* 8 (9–10): 1–34.

van Riel, Raphael, and Robert Van Gulick. 2024. "Scientific Reduction." In *Stanford Encyclopedia of Philosophy*, edited by Edward N. Zalta and Uri Nodelman. Stanford: Metaphysics Research Lab, Stanford University. https://plato.stanford.edu/archives/spr2024/entries/scientific-reduction/.

Weinberg, Steven. 2001. *Facing Up: Science and Its Cultural Adversaries*. Cambridge: Harvard University Press.

Wilson, Mark. 2017. *Physics Avoidance: Essays in Conceptual Strategy*. Oxford: Oxford University Press.

7 Emergence

7.1 A Different Approach

For many reductionists, nothing genuinely new exists in the world. According to this view, all observable objects are simply complex arrangements of elementary particles. While we may find it useful to employ special sciences like chemistry or biology, their use is purely pragmatic. These disciplines don't reveal anything that exists beyond the fundamental domain of physics.

Emergentists take a fundamentally different stance. They acknowledge that higher-level properties and systems depend on lower-level ones, but they reject the idea that higher levels can be completely understood in terms of lower levels. Macroscopic systems exhibit behaviors that cannot be explained by treating them as mere collections of fundamental particles. This philosophical divide has practical implications for how we understand scientific disciplines. For reductionists, the special sciences organize themselves around different scales primarily for convenience. For emergentists, however, this organization reflects the genuine hierarchical structure of nature itself. The fact that chemistry, biology, and other sciences focus on particular levels isn't just a pragmatic choice—it mirrors real organizational features of the natural world (Gillett 2002, 98).

Beyond these broad generalizations, the concept of emergence is notoriously messy. "Nothing perhaps offers more confusion or creates more disagreement when thinking about order, stability, and novelty than talk of emergence" (Bishop, Silberstein, and Pexton 2022, 3). First, what is it that emerges (Seifert 2020, 45)? Properties, events, causes, laws, entities, phenomena? And what is the nature of the dependence of that thing on the base level? Jessica Wilson offers seven possible answers to that question (2021, 14). And how should we understand this notion of levels? Some take it in a robust metaphysical sense: Nature is literally organized into levels with its own distinct entities (which is how the early twentieth-century British

DOI: 10.4324/9781003681199-7

Emergentists thought of it (O'Connor 2014, 208)). Others resist reifying the entities of each level (Ioannidis et al. 2022, 6).

Emergence is the subject of several recent monographs and edited volumes,[1] so choices will have to be made about what to include. Some of the terminology is widely used, such as the distinction between weak and strong emergence. Unfortunately, what is meant by that distinction differs from one source to another. In this chapter, I have chosen what I consider to be the best of the options, but I will not track down all the ways people use the terms. As such, I am imposing some degree of order on an unruly terrain. There is no canonical taxonomy of emergence.

I take no position on what it is that emerges: again, properties, entities, or something else. As such, I freely move from one to the other in this chapter.

Finally, some of the issues introduced here are especially important in the following chapters. I will flag those cases where a fuller explanation will be given later.

7.2 Weak Emergence

While emergence is controversial, one type has become relatively common. *Weak emergence* starts with the benign idea that phenomena at higher levels or scales are qualitatively different from those at lower levels. While H_2O is the chemical formula for water, molecules are not wet. Likewise, the redness of my coffee mug is not due to its being composed of red-colored atoms. For some, that's all weak emergence amounts to: new properties emerging from a base level. "So-called 'weak' emergence just says the obvious thing, that higher-level notions like the fluidity or solidity of a material substance emerge out of the properties of its microscopic constituents" (Carroll 2016). In addition, the use of "emergence" typically signals an epistemic gap between levels. "The traditional definition of emergence found in much of the philosophical literature characterizes it in the following way: A phenomenon is emergent if it cannot be reduced to, explained or predicted from its constituent parts" (Falkenburg and Morrison 2015, 1). Weak emergence is compatible with ontological reduction but is pessimistic about epistemic reduction (Section 6.3). Nature may be too complex for finite beings with limited mental capacities and computational tools to fully bridge theories in different disciplines.

Not every higher-level property counts as weakly emergent. Let's consider some criteria by starting with the failures of reduction (Wayne and Arciszewski 2009, 848). First, if *E* is an emergent property from base level *B*, then *E* resists epistemological reduction in terms of *B*. If *E* could be reduced to *B*, there would be less need to find non-reductionist language to describe their relationship. Talk of emergence was less frequent in the first half of the twentieth century, as optimism about reduction was still running high. Emergence has become more prevalent as the limits of reduction become harder to dismiss.

Second, one cannot predict the behavior of the higher level from knowledge of the base alone. While emergent properties do not violate the laws of the base level, such properties could not have been predicted from an exhaustive knowledge of those laws. "[What] is truly significant about emergent phenomena is that we supposedly *cannot* appeal to microstructures in explaining or predicting these phenomena even though they are constituted by them" (Morrison 2015, 92). One might see this as a consequence of the first criterion. If an epistemological reduction were possible, one presumably could predict what would happen at the higher level from knowledge of the base.

There are disagreements about how the second criterion should be understood. Some weaken "cannot predict" to "have not predicted," leaving the door open to future research. Or one might think of this inability as a human limitation—perhaps a Laplacian demon (Section 2.1.1.2) could make such predictions. In any case, all sides agree there is a limit to the inferences we can currently make about emergent properties from knowledge of the base particles or properties alone.

Third, there is novelty at the higher level—when the properties in *E* are qualitatively distinct from those found in *B*. As physicist Philip Anderson famously put it, "More is different":

> The behavior of large and complex aggregates of elementary particles, it turns out, is not to be understood in terms of a simple extrapolation of the properties of a few particles. Instead, at each level of complexity entirely new properties appear, and the understanding of the new behaviors requires research which I think is as fundamental in its nature as any other...
>
> (Anderson 1972, 393)

Anderson thought condensed matter physics, which includes phenomena like superconductivity, was especially instructive. He goes on to say that novelty shapes the basic concepts of a discipline. "At each stage entirely new laws, concepts, and generalizations are necessary, requiring inspiration and creativity to just as great a degree as in the previous one. Psychology is not applied biology, nor is biology applied chemistry" (Anderson 1972, 393).

Novelty does not negate dependence on a base level. Weak emergence holds that the causal powers of all entities are derived from their base elements. It is therefore consistent with fundamentalism: Fundamental physics is responsible for the causal powers of everything else, even if we can't see how (Bishop, Silberstein, and Pexton 2022, 10).[2]

To avoid confusion, recall what we said in Section 2.4.3 about causal powers. Following Jessica Wilson (2021, 32), I use the term in a metaphysically thin sense: The causal powers of an entity are whatever it can do—whatever causal contribution it makes to a system.[3] Causal powers are not to be understood here as something like metaphysical dispositions or capacities.[4]

These three criteria for weak emergence are primarily epistemological. *We* cannot predict emergent properties from knowledge of the base. And novelty is somewhat subjective, as if a given *E* might seem novel from our point of view, but perhaps not for a more insightful alien species or a highly advanced AI. Philosopher Timothy O'Connor argues instead that weak emergence should not be defined in wholly epistemic terms (2021, sec. 3). For him, the critical difference between strong and weak emergence is whether the causal powers of *E* are fundamental (strong) or provided by the base (weak).

To sum up the three criteria of weak emergence:

1 *Failure of reduction*. Emergent properties (causes, entities, etc.) have not and perhaps cannot be epistemically reduced to lower-level properties. (See Chapter 6 for epistemic reduction.) This means that upper-level phenomena have not/cannot be explained in terms of the lower.
2 *Failure to predict*. Emergent properties (causes, entities, etc.) have not and perhaps cannot be predicted from knowledge of lower-level properties.
3 *Novelty*. Emergent properties (causes, entities, etc.) have a genuinely distinct character as compared to the base level, even though those properties cannot exist without the base level (dependence).

Other types of emergentism accept these criteria as a starting point. They agree that emergence puts limits on bottom-up prediction and involves novel properties. They also think that if emergence is only epistemological, then it isn't very interesting (Wilson 2019, 173). Unless emergence tells us something about reality, why bother with the new terminology?

This debate among emergentists runs parallel to another one in philosophy of science. Section 4.5 briefly considered instrumentalism, the view that science aims only at making successful predictions and technological advances rather than giving a true picture of physical reality. For example, instrumentalists need not believe that electrons exist. They take the idea of electrons to merely be an aid to more practical applications. In contrast, scientific realists believe that science provides an accurate portrayal of reality, or at least approximately so. The success of chemistry and electromagnetism provides evidence that electrons exist. If instrumentalism were correct, says the realist, science would be far less interesting or important.

Many emergentists consider weak emergence to be something like instrumentalism. It isn't all that interesting if it doesn't tell us anything about physical reality (Bishop, Silberstein, and Pexton 2022, 8).

Regarding free will, the scientific challenges are not merely that we cannot figure out how to reconcile physics and libertarianism. Free-will skeptics are making a metaphysical claim. They argue that reality leaves no room for free will and that physics proves it. For those purposes, then, weak emergence is a bit too weak. If emergence is going to help the libertarian, it has to tell us something about nature itself.

7.3 Strong Emergence

Strong emergence takes the three criteria mentioned in the previous section as indicators of emergence but then goes further. It says that novelty should be understood in a metaphysical sense. What emerges is not something new and interesting only from our point of view. Instead, unique causal powers come into being, ones not found at the base level. It isn't merely that we fail to see how base-level properties are responsible for higher-level phenomena. Strong emergence denies that the base level fully determines what happens at the emergent level. In other words, fundamentalism is false. What emerges is just as real as anything in the base level.

Both strong and weak emergence affirm the need for a base level. Emergent properties emerge from something:

> A common baseline assumption of accounts of emergence … is that emergent features minimally nomologically supervene on base features, such that, at least in worlds with the same laws of nature as actually hold, the occurrence of an emergent feature requires the occurrence of some base feature, and (again, in such worlds) any such base feature necessitates the emergent feature.
>
> (Wilson 2021, 141)

Still, not all higher-level properties count as emergent. The shape of a ball is determined by the configuration of molecules that compose the ball. The same goes for the mass of the ball. Mark Bedau calls such properties "resultant," those that can be fully understood in terms of their components (2002, 10).[5] Strongly emergent properties are different. They are not due to the structural properties of the base level.[6] Living organisms, for example, "are not appropriately attributed to even complex configurations of fundamental physical entities; and the same is true for the characteristic features of other special-science entities" (Wilson 2021, 4).

Let's consider two proposed examples. Explaining consciousness has arguably been the primary motivation for strong emergence since the late nineteenth century. Strong emergentists deny that our subjective experience, self-awareness, and ability to perceive the world around us can be fully explained in terms of neural connections and brain chemistry alone. This idea traces back to the "British Emergentists" starting with John Stuart Mill, continuing through C. D. Broad (O'Connor 2021, sec. 1) and neuroscientist Roger Sperry (Sperry 1969), to today (Chalmers 2003; O'Connor 2014; O'Connor and Wong 2005).

If consciousness is emergent, there is no such thing as disembodied conscious awareness, at least not in any nomologically possible world (i.e., a world with the same laws of nature as our own).[7] Consciousness requires at the very least a brain (or an equivalent base). Not even God could make a rock conscious on this view. Contrast this with interactive dualism discussed in Section 3.5. The most common form of dualism takes the immaterial mind to be a different substance than the material body. A perennial question for dualism is why one must be paired with the other (Bailey, Rasmussen, and Horn 2011). Why can't a rock have a mind and be conscious? Emergentism does not face the pairing problem, given the mind's intrinsic dependence on a base level.

The second example is entanglement, a well-known oddity of quantum mechanics (Humphreys 1997, 15–16; Silberstein and McGeever 1999, 187–89). When particles interact, they can become entangled—a technical term. When that happens, the individual particles can no longer be described as having their own states. The entangled system must be treated as an irreducible whole, as Tim Maudlin explains:

> [Quantum] holism ought to give some metaphysicians pause... [One] popular "Humean" thesis holds that all global matters of fact supervene on local matters of fact... Once the local facts have been determined, all one needs to do is distribute them throughout all of space-time to generate a complete physical universe. Quantum holism suggests that our world just doesn't work like that. *The whole has physical states that are not determined by, or derivable from, the states of the parts. Indeed, in many cases ... parts fail to have physical states at all.*
>
> (Maudlin 1998, 58–59, italics added)

While reductionists hold out hope that examples of emergence might one day be fully explained in terms of their respective bases, this is simply impossible for entangled systems. Quantum mechanics is already part of fundamental physics, and entanglement has a precise mathematical description.

One of the ways in which strong emergence is metaphysically thick has to do with causal efficacy (O'Connor 2021, sec. 2.2.3). Strongly emergent entities have causal powers that are found neither in nor derived from the causal powers of their base (Bedau 2002, 10). Such powers are novel not merely in the sense of being unexpected but also in that they can do things that cannot be done by collections of base-level entities. Gillet calls this Higher Causal Efficacy (HCE). Our own mental states are a key example:

> HCE is strongly supported by our everyday experiences of the apparent causal efficacy of our own beliefs, desires, and other mental states, as well as by the findings of the special sciences. Given the character of causal efficacy, HCE implies that realized and other higher level properties actually contribute causal powers to individuals, or otherwise determine their contribution.
>
> (Gillett 2002, 99)

Emergent properties bring new causal capacities into being.

One of these capacities gets a lot of attention, both positive and negative. While causation is typically thought of in bottom-up terms (e.g., the motion of molecules causing a change in pressure) or as an intra-level relation (e.g., a change in pressure causes a change in volume), strongly emergent properties can generate downward causation. Emergent properties/states produce effects at lower levels. Mental states are again a convenient example. One's intention to reach for a coffee mug has the effect of moving the atoms in one's hand. The causal chain runs from a high-level mental state to an effect at a lower level. O'Connor argues we need to broaden the way we typically think of causation, replacing "the picture of mostly horizontal, or same-level, causal patterns within each level with one in which there is a complex web of myriad upward and downward causal influences that jointly determine the system's evolution through time ..." (2014, 208). Others, like Gillet, argue that causation only applies to intra-level events and that we should use a different term to talk about emergent entities determining or constraining what goes on at lower levels (2016, 64).[8] (We will take up the question of downward causation again in Section 9.2.3.)

One way to avoid questions about causation is to talk instead about autonomy: Strongly emergent properties are not wholly dictated by lower levels. Indeed, as we will see more fully in Chapter 8, what goes on at lower levels is sometimes irrelevant to the behavior of emergent systems. This is a striking contrast with most types of reduction, which assume that the lowest level is ultimately running the show whether or not we know how. Autonomy can also be used to distinguish strong from weak emergence. The latter affirms a kind of dependence captured by the principle of supervenience (Section 6.4.4). Strong emergence denies that supervenience always holds (Silberstein and McGeever 1999, 186).

> Emergent properties do not supervene on basal properties even though they are caused and sustained by basal properties, because there can be

> differences in emergent properties without a change in basal properties due to the novel causal powers of the emergent properties.
>
> (Wong 2019, 182)

Robert Batterman argues that autonomy is the most important attribute on this list, although he has come to believe that the reduction/emergence dichotomy is not helpful (2021, 27–28).

Strong emergence conflicts with at least two other reductionist principles. The first is the causal closure of physics (Wilson 2021, 46–50). Emergent properties have a causal efficacy that is not derived from the causal powers of fundamental physics, and yet HCE is just as real as causation at the base level. If strongly emergent properties were responsible for changes to physical systems, that would violate causal closure.

The second is physicalism.[9] Strong emergence entails that entities, properties, or causal powers that are not merely resultant (in Bedau's sense) exist at higher levels. Epistemological and ontological reduction fail in such cases. There is more to reality than physics can describe, even in principle.

Summing up strong emergence in terms of the three criteria mentioned above:

1 *Failure of reduction*. Emergent properties (causes, entities, etc.) cannot be epistemically or ontologically reduced. Not all upper-level phenomena can be explained in terms of the lower.
2 *Failure to predict*. Emergent properties (causes, entities, etc.) cannot be predicted from knowledge of lower-level properties.
3 *Novelty*. Emergent properties (causes, entities, etc.) have genuinely distinct causal powers, including downward causation, as compared to the base level, even though those properties cannot exist without the base level (dependence).

These three are necessary but not sufficient for strong emergence.

It should now be clear why strong emergence is controversial. The rejection of physicalism bothers both philosophers and scientists. "Most physicists are confident strong emergence doesn't exist. The reason is not only that there isn't any known example for it but that, more importantly, if there was an example it would be incompatible with that [*sic*] they already know" (Hossenfelder 2019, 89). Let's spell out what Hossenfelder has in mind.

7.3.1 *No Empirical Support*

Even if strong emergence were a coherent position, its critics say we have yet to detect it (McLaughlin 1992, 55). "All the evidence today suggests that strong emergence is scientifically irrelevant" (Bedau 2002, 11).

> In physics the only type of emergence we have is weak emergence. With strong emergence philosophers refer to the hypothetical possibility that a system with many constituents displays a novel behavior which cannot be derived from the properties and the interactions of the constituents. While this is logically possible, there is not a single known example for this in the real world…
>
> (Hossenfelder 2020)

We considered two examples of strong emergence earlier. What about those? Consciousness is arguably the most challenging scientific and philosophical question currently being pursued. But it fails to provide a clear picture of what it is supposedly an example of and offers no guidance for finding or understanding other examples. If strong emergence is somewhat mysterious, appealing to consciousness exchanges one mystery for another.

While quantum entanglement is better understood, it isn't an example of *interlevel* emergence (Wilson 2021, 123, fn3). An entangled system must be mathematically described as a whole rather than an ensemble of individual particles. But the new system exists at the same level as the particles. It does not exhibit causal powers that are new and different from those of its constituents. Entanglement even fails to pass the first criterion of *weak* emergence since the behavior of the whole can be derived from knowledge of its parts (assuming that this knowledge would include the fact that they were entangled (Hossenfelder 2019, 92)).

In short, there are no uncontroversial examples of strong emergence. If consciousness and entanglement are the best of the bunch, that does not bode well for the program. One might think, however, that this negative assessment misunderstands how philosophy works. Metaphysics is seldom based on new discoveries. It often concerns the best way to understand what we already know. So while consciousness is mysterious, thinking of it in terms of strong emergence might provide a new way of approaching the mystery. Likewise when it comes to the failures of reduction. Perhaps a strong emergence perspective can help us understand why reduction sometimes fails.

Well, perhaps. The burden still falls on the strong emergentist to make the case.

7.3.2 Causal Closure

As we saw in Chapter 5, many philosophers and scientists accept CCP, but strong emergence violates closure in two ways. First, HCE posits irreducible causes above the level of physics. When physicalists use causal talk for higher-level phenomena, they often do so for pragmatic reasons. While a physicalist might say, "A lack of rain caused the mosquito population to drop," but they actually deny real biological causation. All such language is merely shorthand for the unknown, complex interactions in physics that are ultimately responsible for biological events. Strong emergence, however, takes (at least some) higher-level causation to be ontologically on par with that in physics. Second, downward causation would involve higher-level causes that sometimes directly bring about physical effects—a clear violation of CCP.

The strong emergentist has a better response available to this objection: CCP isn't part of physics (as Wilson rightly observes (2021, 124) and as I argued in Section 5.4). While strong emergence is in conflict with physicalism and CCP, these are philosophical, not scientific positions. They are not, as Hossenfelder claims, part of what physicists "already know" (at least not qua physicists). When strong emergence rejects CCP it is, as they say, a feature not a bug. It is part of what their position is asserting, not a problem to be overcome.

7.3.3 Force Laws

Contemporary physics recognizes four fundamental forces (gravity, electromagnetism, plus the strong and weak nuclear forces). Macroscopic forces are either not really forces (like friction) or are reducible to the four fundamental forces. The force my fingers exert on the keyboard, for example, results from the electromagnetic repulsion of electrons in the constituent atoms of both.[10]

Now the problem. An emergent cause that brings about a change in a physical system seems to involve a force. But forces that act from higher to lower levels are not among the four fundamental forces (nor are they reducible to those four forces).

Carroll puts the objection in mathematical terms (2011). While he's arguing against interactive dualism, what he says applies to strong emergence. Suppose that consciousness and mental states (beliefs, desires, intentions, etc.) are strongly emergent. If one's intention to write a note causes the electrons in one's hand to move, that would require some sort of influence that moves those electrons. The behavior of fermions, including electrons, is described by the Dirac equation. Paul Dirac's Nobel Prize in Physics in 1933 was based in part on the discovery of this equation. If mental states were influencing the electrons in one's hand, says Carroll, then the Dirac equation would be wrong. The equation fails to include a term for this new influence. But since the equation has been empirically verified, it appears that there is no such influence.

The example is a bit oversimplified. Strong emergentists don't claim that mental states move electrons. Still, Carroll could make the argument using other examples. His point is that there is no place within the laws of physics for downward causation. But does that mean that physicists have not found evidence for downward causation, and so have not included it in the laws? If so, Carroll's objection is just another example of the "no empirical support" problem (Section 7.3.1): If physicists haven't detected downward causation in their labs, perhaps they're looking in the wrong place. As Brian Pitts puts it in his reply to Carroll, "[Evidence] for anomalous events needs to be sought where and when such events allegedly happen(ed), not where one finds it convenient to look …" (2020, 1940).

But perhaps Carroll is making a stronger claim. Instead of merely lacking evidence, perhaps physics *forbids* new forces.

> Strong emergence is incompatible with what we already know about the laws of nature. That's because if you think that strong emergence can really happen, then this necessarily implies that there will be objects in this world whose behavior is in conflict with the standard model of particle physics. If that wasn't so, then really it wouldn't be strong emergence.
>
> (Hossenfelder 2020)

Hossenfelder believes that events at the lowest level determine higher-level phenomena and not vice versa. Causal arrows running the other way are forbidden.

But suppose we grant that strong emergence involves forces.[11] Classical physics doesn't forbid new forces. The laws of mechanics recognize different types of force and would have no problem accommodating others.[12] And the laws don't care where forces originate. Newton's theory of gravity seemed to imply action-at-a-distance whereby one object could gravitationally attract another without any physical connection between them. While physicists no longer believe there are such non-local forces, the point is that the laws of mechanics don't care one way or the other. A force is a force regardless of how it arises. An analogous story can be told in terms of quantum mechanics, although forces play a lesser role there (McLaughlin 1992, 53–54).

Current attitudes toward downward causation are like those regarding action-at-a-distance: Based on the evidence, physics doesn't need it. But that doesn't mean that physics forbids it. If sufficient evidence were found, then the equations would be changed to reflect it (although as Pitts remarks, that sort of evidence isn't likely to be found in a particle accelerator (2020, 1939)). Wilson makes a similar point concerning conservation laws:

> [Scientific] theorizing itself provides a blueprint for how conservation laws might enter into the warranted posit of Strongly emergent configurational interactions and associated powers, laws, and features. In the

> 1930's, the law of conservation of mass-energy appeared to be violated in nuclear β-decay interactions. Rather than accept the apparent violation as genuine, physicists posited a new fundamental interaction—the weak nuclear interaction—as carrying away the missing energy... [The strong emergentist] posit of fundamental configurational forces/interactions and associated powers, which come into play only at certain levels of comparatively complex organization, is compatible with scientific practice...
>
> (Wilson 2021, 123)

There is simply no good argument that downward causation is incompatible with physics.

7.3.4 *Too Weird*

The more conservative argument against strong emergence is that it just doesn't fit well with contemporary science. Both the small number of examples offered and the fact that those examples are not well-understood make emergence feel "kooky" (McLaughlin 1992, 55). Proponents of emergence recognize this:

> If the only cases of [strong] emergence ... are quantum weirdness, perhaps the emergence of life, and experience, this would be a very strange disunified world and ontological emergence would be a topic with a very limited and relatively rare subject matter. However, if the only choice is between the very common cases of weak or epistemic emergence on the one hand and spooky strong emergence on the other ... , then we understand why the preferred choice by default, is weak emergence.
>
> (Bishop, Silberstein, and Pexton 2022, xix)

For many, strong emergence is a metaphysical bridge too far.

The challenge, then, is to formulate a view that captures the metaphysical significance of emergence—its reality and causal relevance—without inheriting the "spooky" baggage of strong emergence. The next two positions we will consider, contextual emergence and weak metaphysical emergence, can be seen as sophisticated attempts to chart precisely this middle path.

7.4 Contextual Emergence

Reduction, even for non-reductive physicalists, takes nature to operate from the bottom up: Lower levels somehow or other provide sufficient conditions for events at the higher levels. The higher is determined by the lower. Bishop, Silberstein, and Pexton (2022) argue that's not always the case. Higher-level environmental or contextual conditions also typically shape the behavior of systems.[13] Consider chemical isomers (2022, 118–19). These are molecules with the same constituent atoms but different bonding patterns between those atoms. For example, C_3H_4 is the formula for two distinct chemical compounds: allene and methyl acetylene. Three things determine which isomer will form. One is temperature. Higher temperatures favor more thermodynamically stable compounds. Another is pressure, especially in gases. A third is the polarity of the solvent in which the isomer forms. The point: the properties of the compound are not wholly determined by its parts.[14] Higher-level environmental factors play a crucial role, something that reductionism fails to capture:

> Contextual emergence challenges [the] "foundationalist" or hierarchical picture of reality. Contextual emergence is a rejection of the very idea of ontological autonomy *at any scale* and thus a rejection of the idea that determination and hence explanation, must strictly be "bottom-up," at least in principle.
>
> (Bishop, Silberstein, and Pexton 2022, 10, italics added)

The emphasis highlights a surprising claim. Not even the fundamental level, if there is one, has properties wholly independent of context.

Another example is Rayleigh-Bénard convection (Bishop, Silberstein, and Pexton 2022, 41–42). When a fluid is heated from below, convection cells can form where the warmer fluid at the bottom rises to the top while the cooler fluid sinks. This observable process constrains how the molecules in the fluid behave. Yet this behavior is not wholly due to the interaction of those molecules. Rayleigh-Bénard cells cannot form without the heating element, a macroscopic feature of the system. Again, the point: large-scale context matters, not just the behavior of molecules.

Some of the conditions needed for emergence—what they call "stability conditions"—are causal and mechanical, like the plates in the convection examples. Others are acausal, like conservation laws and least action principles. Some are abstract properties of dimension and the geometry of mathematical spaces. An example of this last type is stable planetary orbits, which require three-dimensional space. In four dimensions, regular orbits that endure small perturbations would be impossible (Bishop, Silberstein, and Pexton 2022, 29). Note that the geometry of space is not part of the system. It is the context in which the system exists. Three dimensions are a necessary condition for stable orbits that cannot be reduced to the system's constituents (even in principle). In terms of the three criteria for emergence, this shows why contextually emergent properties are frequently inexplicable or unpredictable, even with complete knowledge of the underlying lower-level components. Stability conditions are not typically found at some lower level. Nonetheless, those conditions are as real or "fundamental" as anything in elementary physics.

While all types of emergence involve novelty, that criterion plays a lesser role in contextual emergence. It emphasizes instead how stability conditions shape the "possibility space" of a system. A possibility space represents all potential states or behaviors of a system (similar to the phase spaces discussed in Section 3.2). To illustrate this concept, consider a baseball. One point in its possibility space represents the ball orbiting the Earth—a possibility that will likely never be actualized. In Newtonian mechanics, the ball could theoretically travel at the speed of light, but special relativity closes off this part of the possibility space, rendering it inaccessible to material objects. Stability conditions can also create access to previously closed areas of possibility space. For instance, lasers do not occur naturally. Their stability conditions require a resonance cavity where atoms can be electrically stimulated and isolated from their environment (Bishop, Silberstein, and Pexton 2022, 152–53). When these conditions are met, the area of possibility space representing coherent light becomes accessible. This type of light has always been physically possible, but without the proper context, it cannot become a reality.

By reframing the discussion around stability conditions and possibility space, Bishop et al. offer a unique perspective on emergence. Unlike cases of weak emergence, examples of contextual emergence are not grounded in what we do not know. There is no mystery about how Rayleigh-Bénard cells form. It is obvious why the behavior of the cells cannot be reduced to the behavior of their constituent molecules. Higher-level factors are necessary conditions. Unlike strong emergence, contextual emergence is "not rare or exotic" (Bishop, Silberstein, and Pexton 2022, 8). Once you have the idea, there is no difficulty finding examples.

There is, however, one major objection that needs to be addressed. As we've seen, contextual emergence requires stability conditions. Reductionists claim that those conditions can themselves be described in base-level terms (Wilson 2021, 124, n3). It is fundamental physics that ultimately governs the heating element in Rayleigh-Bénard convection, they say, not something that is intrinsically macroscopic. The same goes for pressure and temperature in isomer formation. Whatever environmental conditions Bishop et al. point to can (in principle!) be reduced to the properties of a base level. I believe this is the most important objection to all forms of emergence. It will be addressed in Chapter 8.

While contextual emergence emphasizes stability conditions, Jessica Wilson offers another intermediate position between weak and strong emergence that focuses on degrees of freedom (DOF). Her account provides a way to understand emergence without abandoning physicalism.

7.5 Weak Metaphysical Emergence

Wilson begins by placing the burden of proof on the reductionist. Anyone familiar with the hierarchical structure of the natural sciences must admit that emergent properties *seem* to be found at higher levels. If reductionists dispute this, Wilson argues, it is up to them to show otherwise (2021, 36). Her "weak metaphysical emergence" is weak in that, unlike strong emergence, it does not conflict with physicalism. No expanded ontology of novel causal powers is needed. It is metaphysical in that it is not merely epistemic. Emergence is a real phenomenon. Scientists and philosophers of science need it to correctly describe nature.

She summarizes her view this way (which will need some unpacking):

> What it is for token feature *S* to be Weakly metaphysically emergent from token feature *P* on a given occasion is for it to be the case, on that occasion, (i) that *S* cotemporally materially depends on *P*, and (ii) that *S* has a non-empty proper subset of the token powers had by *P*.
>
> (Wilson 2021, 75)

While heavy on jargon, the ideas are not complicated. To start, she does not promote a narrow view of what sorts of things are emergent. *S* might be a property, state, behavior, or something else (2021, 1). A "token" is just one instance of some property. Tokens are contrasted with types. C sharp is a type of musical note. Striking the proper key on a piano produces one token of that type. "Powers" for Wilson are just what things are able to do. One distinctive aspect of her account is that *S*'s causal powers are a subset of *P*'s causal powers. There are no new causal powers at higher levels, which is why weak metaphysical emergence is compatible with physicalism. The space of physically possible behaviors is defined by *P*. Whatever novelty is found in *S* is not fundamental.

One problem with this is that physicists don't talk about causal powers. She bridges this gap by referring instead to "degrees of freedom." A system's DOF are the different, independent ways it can move or change. Consider an ideal rod (incompressible and unbendable). The rod can travel along three different axes in space. It can also rotate around each of those three, giving the rod six DOF. A given system might have reduced, constrained, or eliminated DOF (2021, 178–79). A bowling ball in the gutter can only roll forward or backward. The same ball on a gymnasium floor can roll in any combination of two different directions. The ball in the gutter has a reduced number of DOF. Its freedom has been constrained if barriers are set up on the floor so the ball cannot reach some areas. Elimination happens when the composite no longer depends on all the DOF found at lower levels. For example, quantum mechanical particles have a property known as "spin." An electron might be spin up or spin down with respect to some axis. A rod containing those same electrons does not have quantum spin. That DOF has been eliminated.

Wilson argues that the properties of composite entities like a rod cannot be identified with the properties of any collection of fundamental particles. Those particles have spin. The rod does not. Lower-level laws are in a sense blind to a composite entity *E* in which some DOF have been eliminated. Those laws need all such DOF to work:

> Since *E*'s characterization, by assumption, fails to include (DOF containing) any information about quantum spin, the lower-level physical laws will not be able to take an entity such as *E* as input. *E*, and more generally any goings-on whose DOF are eliminated vis-à-vis their lower-level physical realizers, cannot be identified with any lower-level goings-on, since their specifications fail to include all the DOF needed for the lower-level physical laws to operate. This much shows that an entity *E*'s satisfaction of the conditions in *DOF-based Weak emergence* suffices to ensure that *E* is ontologically autonomous—that is, distinct—from any lower-level goings-on.
>
> (Wilson 2021, 181)

While Wilson talks about emergent entities, it's their properties that are doing the work. The rod is not an entity that quantum field theory and the Standard Model can straightforwardly deal with because it lacks the properties needed for the equations to work. Consider an analogy. Statics is a branch of classical mechanics that studies the effects of forces on unmoving rigid bodies. Civil engineers use it to, among other things, analyze bridge designs. Water flowing through a pipe is not a proper system from a statics point of view. It does not have the properties needed for its laws to apply. Likewise for a composite entity *E* from a quantum mechanical point of view. *E* itself lacks the properties needed for fundamental laws to operate.

Wilson gives three examples of weak metaphysical emergence. The first is cellular automata, like the Game of Life (2021, 167). The computerized game is a two-dimensional landscape divided into individual cells, much like pixels on a computer screen. Unlike standard video games, the Game of Life only specifies a small set of rules defining when each cell is "on" based on the state of its neighbors. The game begins with a random distribution of on and off cells. As the game progresses, small "creatures" known as gliders spontaneously develop. At the level of the program, these are simply cells turning on or off according to their local rules. But to observers attuned to pattern recognition like us, they appear to be entities that move across the board. It is important to note that gliders were not intentionally programmed into the game.

For the program to determine whether a given cell is on requires ten pieces of information about that cell and its immediate neighbors. These constitute DOFs in Wilson's account. The cells composing a glider only have two DOFs corresponding to their position in the two-dimensional grid. DOFs at the base level are eliminated at the game level. Finally, gliders themselves are not something that the algorithm controlling the cells recognizes since it has too few DOFs, and yet they clearly depend on the algorithm to come into being. Gliders meet the criteria for weak metaphysical emergence.[15] If the "existence" of a glider seems a bit too insubstantial, flocking birds (e.g., starlings) provide a kind of real-life analog to cellular automata (Wilson 2021, 188–89).

An example that will be important in Chapter 8 has to do with complex systems and phase transitions (Wilson 2021, 185–87). These exhibit *universality*: when a property or behavior is found in systems with very different physical compositions. Each element on the periodic table exists in the three common states of solid, liquid, and gas. The change from one state to another is a phase transition. The idea applies more broadly, however. Magnets undergo phase transitions (e.g., ferromagnetic to paramagnetic). So do superconductors. Robert Batterman has argued that we need to consider how systems with completely different physical compositions display this universal behavior (Batterman 2000, 2009). The explanation cannot be reductive since the micro-constituents are heterogeneous. Exhaustive knowledge of the components does not explain universal behavior.

Wilson notes three features of Batterman's account that are relevant to weak metaphysical emergence (2021, 175). One is the "elimination of micro-level degrees of freedom (DOF)," which we have discussed. Upper levels no longer have all the DOFs found in the base level. Another feature is "stability of certain behavior under perturbation." If you nudge the moving pendulum of a grandfather clock, that's a kind of perturbation. If the nudge isn't too strong, the pendulum will eventually return to its regular, periodic behavior. Some readers might have had a standup, inflatable punching bag as a toy (Figure 7.1). After the bag has been punched, the weight in the bottom pulls it upright again. Both systems exhibit stability under perturbation. The final feature is universality. Emergent phenomena are found across systems with different micro-constituents and therefore cannot be explained in terms of causal interactions at lower levels.

Wilson argues that complex systems—especially chaotic, nonlinear systems—and phase transitions exhibit all three features, providing empirical support for weak metaphysical emergence (2021, 187). She believes their most important feature vis-à-vis emergence is the elimination of DOFs. As I'll argue later, however, it's universality.

To summarize, weak metaphysical emergence occurs when higher-level features depend on lower-level ones but possess only a subset of their

Figure 7.1 Punching bag.

Source: Artwork by Ashley McNutt-Greene. Used with permission.

causal powers or DOF. This elimination of DOF renders emergent entities autonomous from lower levels without requiring new causal powers. Unlike strong emergence, this view is compatible with physicalism while still granting higher-level phenomena a kind of ontological independence.

7.6 Where to from Here?

We have considered four types of emergence in this chapter. The first two represent the extremes. Scientists referring to emergence typically mean it in an epistemic sense, i.e., weak emergence. That there are qualitative differences between levels is uncontroversial. New terminology is needed when we realize that a property or entity has not been (or perhaps cannot be) epistemically reduced. Weak emergence fits that need.

At the other extreme is strong emergence, with an expanded ontology of autonomous properties and novel causal powers. Strong emergence is an exciting idea insofar as it posits new, fundamental features of physical reality. But with a short list of examples, all of which are controversial, it has not garnered much support.

Two other positions fall somewhere in between. Contextual emergence challenges the reductionist perspective that the most fundamental level somehow determines everything. Phenomena at each level are instead dependent on their environments and sometimes history, which are themselves defined in terms of higher levels. Lastly, weak metaphysical emergence finds a kind of autonomy at higher levels by eliminating DOFs. Emergent entities are no longer "recognized" by lower-level laws since they lack the properties on which those laws operate.

These latter two positions will help orient what is to come, although my view does not fit neatly into either. In broad terms, I'll argue for a kind of metaphysical emergence. It differs from Wilson's view in that emergence involves adding or accessing new DOFs, not just eliminating them. It differs from strong emergence in that it is far more common and somewhat less mysterious. Saving the details for later, for now let's call it *moderate metaphysical emergence*.

As a reminder, a viable account of emergence is not the final destination. We are still looking to fulfill the "hard way" for libertarian freedom to be possible considering everything we know about physics.[16] Emergence will be a crucial part of that story.

Notes

1 Among the monographs, see Wilson (2021), Bishop, Silberstein, and Pexton (2022), and Bishop (2024). The best recent edited volume is Gibb, Hendry, and Lancaster (2019), but also see Bedau and Humphreys (2008).

2 Fundamentalism was mentioned in the preface. Bishop et al. call it "foundationalism." It is closely related to physicalism, which I have used more frequently to this point. Strictly speaking, physicalism is a metaphysical claim: all that exists is physical stuff. Fundamentalism adds to physicalism the primacy of the lowest level of reality, especially when it comes to scientific explanations. Reductive physicalists are fundamentalists. Non-reductive physicalists (Section 6.6.4) might or might not accept fundamentalism.

3 More from Wilson:

> That features are associated with actual or potential causal contributions ("powers") reflects the uncontroversial fact that what entities do (can do, relative to the same laws of nature) depends on how they are (what features they have). So, for example, a magnet attracts nearby pins in virtue of being magnetic, not massy; a magnet falls to the ground when dropped in virtue of being massy, not magnetic (Wilson 2021, 31).

4 For a metaphysically thick (non-Humean) alternative, O'Connor and Wong argue for a neo-Aristotelian understanding of causal powers (O'Connor 2021, sec. 4.2.2; O'Connor and Wong 2005, 665–68).

5 This language goes back to C. D. Broad ([1925] 2002).

6 Carl Gillet makes the same distinction using different terms. An entity whose abilities comes from its parts is a "collective." Those that have their own determinative abilities are "composed" (Gillett 2016, 276). A pile of sand is a collective. Its shape is determined by the arrangement of its constitutive molecules. A conscious person, Gillet argues, is composed of cells, but has causal powers in addition to those provided by its cells.
7 At least for creatures that are normally embodied. Emergence would not apply to God, for example.
8 Gillet prefers "machresis," a kind of downward determination that is explicitly not causation (2016, 207; 2019, 101).
9 Non-reductive physicalism (Section 6.6.4) denies full-blown epistemological reduction. Simply put, physicalism is the correct ontology, although we will never be able to show how the special sciences can be reduced to physics.
10 The Pauli Exclusion Principle also plays a role.
11 Causation does not always involve forces. A lack of rain does cause mosquito populations to drop. Nothing in that causal claim entails a force moving insects around.
12 Body forces like gravity act within each part of an object. Contact forces, like pushing on a door, only act along a surface. The laws can incorporate any number of either, although the resulting equations might not be solvable.
13 Also see Bishop (2012), Silberstein (2017), Bishop and Ellis (2020), and Bishop (2024).
14 Or consider carbon. Diamonds are electric insulators, yet graphene is a conductor (Bishop 2024, sec. 5.3). Both are composed of carbon.
15 Bishop et al. believe the Game of Life is too limited to be of interest since the outcome of a given "game" is determined by the algorithm controlling the cells plus the initial conditions (Bishop, Silberstein, and Pexton 2022, 290). More generally, they believe that Wilson's weak metaphysical emergence is too weak since it does not break free of physicalism.
16 The "easy way" is interactive dualism (Chapter 5).

Works Cited

Anderson, Philip W. 1972. "More Is Different: Broken Symmetry and the Nature of the Hierarchical Structure of Science." *Science* 177 (4047): 393–96. https://doi.org/10.1126/science.177.4047.393.

Bailey, Andrew M., Joshua Rasmussen, and Luke Van Horn. 2011. "No Pairing Problem." *Philosophical Studies* 154 (3): 349–60. https://doi.org/10.1007/s11098-010-9555-7.

Batterman, Robert W. 2000. "Multiple Realizability and Universality." *The British Journal for the Philosophy of Science* 51 (1): 115–45.

Batterman, Robert W. 2009. "Emergence in Physics." In *Routledge Encyclopedia of Philosophy*, edited by Edward Craig. London: Routledge.

Batterman, Robert W. 2021. *The Middle Way: A Non-Fundamental Approach to Many-Body Physics*. Oxford: Oxford University Press.

Bedau, Mark A. 2002. "Downward Causation and the Autonomy of Weak Emergence." *Principia: Revista Internacional de Epistemologia* 6 (1): 5–50.

Bedau, Mark A., and Paul Humphreys, eds. 2008. "Preface." In *Emergence: Contemporary Readings in Philosophy and Science*, i–viii. Cambridge: MIT Press.

Bishop, Robert C. 2012. "Fluid Convection, Constraint and Causation." *Interface Focus* 2 (1): 4–12. https://doi.org/10.1098/rsfs.2011.0065.

Bishop, Robert C. 2024. *The Physics of Emergence*. 2nd ed. San Rafael: IOP Publishing. https://doi.org/10.1088/978-0-7503-6367-9.

Bishop, Robert C., and George F. R. Ellis. 2020. "Contextual Emergence of Physical Properties." *Foundations of Physics* 50 (5): 481–510. https://doi.org/10.1007/s10701-020-00333-9.

Bishop, Robert C., Michael Silberstein, and Mark Pexton. 2022. *Emergence in Context: A Treatise in Twenty-First Century Natural Philosophy*. Oxford: Oxford University Press.

Broad, C. D. (1925) 2002. *The Mind and Its Place in Nature*. Reprint. Philosophy of Mind and Language. Vol. 3. London: Routledge.

Carroll, Sean. 2011. "Physics and the Immortality of the Soul." *Scientific American Blog Network*. May 23, 2011. https://blogs.scientificamerican.com/guest-blog/physics-and-the-immortality-of-the-soul/.

Carroll, Sean. 2016. "Consciousness and Downward Causation." *Preposterous Universe* (blog). September 8, 2016. www.preposterousuniverse.com/blog/2016/09/08/consciousness-and-downward-causation/.

Chalmers, David. 2003. "Consciousness and Its Place in Nature." In *Blackwell Guide to the Philosophy of Mind*, edited by Stephen P. Stich and Ted A. Warfield, 102–42. Oxford: Blackwell.

Falkenburg, Brigitte, and Margaret Morrison. 2015. "Introduction." In *Why More Is Different*, edited by Brigitte Falkenburg and Margaret Morrison, 1–10. The Frontiers Collection. Berlin: Springer. https://doi.org/10.1007/978-3-662-43911-1_6.

Gibb, S. C., Robin Findlay Hendry, and Tom Lancaster, eds. 2019. *The Routledge Handbook of Emergence*. Routledge Handbooks Philosophy. London: Routledge.

Gillett, Carl. 2002. "The Varieties of Emergence: Their Purposes, Obligations and Importance." *Grazer Philosophische Studien* 65 (1): 95–121. https://doi.org/10.1163/18756735-90000795.

Gillett, Carl. 2016. *Reduction and Emergence in Science and Philosophy*. Cambridge: Cambridge University Press. https://doi.org/10.1017/CBO9781139871716.

Gillett, Carl. 2019. "Emergence, Downward Causation and Its Alternatives: Critically Surveying a Foundational Issue." In *The Routledge Handbook of Emergence*, edited by Sophie Gibb, Robin Findlay Hendry, and Tom Lancaster, 99–110. Routledge Handbooks in Philosophy. London: Routledge.

Hossenfelder, Sabine. 2019. "The Case for Strong Emergence." In *What Is Fundamental?*, edited by Anthony Aguirre, Brendan Foster, and Zeeya Merali, 85–94. The Frontiers Collection. Cham: Springer International Publishing. https://doi.org/10.1007/978-3-030-11301-8_9.

Hossenfelder, Sabine. 2020. "What Is Emergence? What Does 'Emergent' Mean?" *Backreaction* (blog). April 3, 2020. http://backreaction.blogspot.com/2020/04/what-is-emergence-what-means-emergent.html.

Humphreys, Paul. 1997. "How Properties Emerge." *Philosophy of Science* 64 (1): 1–17.

Ioannidis, Stavros, Gal Vishne, Meir Hemmo, and Orly Shenker. 2022. "Introduction." In *Levels of Reality in Science and Philosophy*, edited by Stavros Ioannidis, Gal Vishne, Meir Hemmo, and Orly Shenker, 1–10. Jerusalem Studies in Philosophy and History of Science. Cham: Springer International Publishing. https://doi.org/10.1007/978-3-030-99425-9_1.

Maudlin, Tim. 1998. "Part and Whole in Quantum Mechanics." In *Interpreting Bodies: Classical and Quantum Objects in Modern Physics*, edited by E. Castellani, 46–60. Princeton: Princeton University Press.

McLaughlin, Brian. 1992. "The Rise and Fall of British Emergentism." In *Emergence or Reduction? Essays on the Prospects of Nonreductive Physicalism*, edited by Ansgar Beckermann, Hans Flohr, and Jaegwon Kim, 49–93. Foundations of Communication and Cognition. Berlin: W. de Gruyter.

Morrison, Margaret. 2015. "Why Is More Different?" In *Why More Is Different*, edited by Brigitte Falkenburg and Margaret Morrison, 91–114. The Frontiers Collection. Berlin: Springer. https://doi.org/10.1007/978-3-662-43911-1_6.

O'Connor, Timothy. 2014. "Philosophical Implications of Emergence." In *The Routledge Companion to Religion and Science*, edited by James W. Haag, 206–12. London: Routledge.

O'Connor, Timothy. 2021. "Emergent Properties." In *Stanford Encyclopedia of Philosophy*, edited by Edward N. Zalta. Stanford: Metaphysics Research Lab, Stanford University. https://plato.stanford.edu/entries/properties-emergent/.

O'Connor, Timothy, and Hong Yu Wong. 2005. "The Metaphysics of Emergence." *Noûs* 39 (4): 658–78.

Pitts, J. Brian. 2020. "General Relativity, Mental Causation, and Energy Conservation." *Erkenntnis*, 1931–73. https://doi.org/10.1007/s10670-020-00284-7.

Seifert, Vanessa A. 2020. "The Strong Emergence of Molecular Structure." *European Journal for Philosophy of Science* 10 (3): 45. https://doi.org/10.1007/s13194-020-00308-7.

Silberstein, Michael. 2017. "Strong Emergence No, Contextual Emergence Yes." *Philosophica* 91 (January):145–92.

Silberstein, Michael, and John McGeever. 1999. "The Search for Ontological Emergence." *The Philosophical Quarterly* 49 (195): 182–200. https://doi.org/10.1111/1467-9213.00136.

Sperry, R. W. 1969. "A Modified Concept of Consciousness." *Psychological Review* 76 (6): 532–36. https://doi.org/10.1037/h0028156.

Wayne, Andrew, and Michal Arciszewski. 2009. "Emergence in Physics." *Philosophy Compass* 4 (5): 846–58. https://doi.org/10.1111/j.1747-9991.2009.00239.x.

Wilson, Jessica M. 2019. "Between Scientism and Abstractionism in the Metaphysics of Emergence." In *The Routledge Handbook of Emergence*, edited by Sophie Gibb, Robin Findlay Hendry, and Tom Lancaster, 157–76. London: Routledge. https://doi.org/10.4324/9781315675213-13.

Wilson, Jessica M. 2021. *Metaphysical Emergence*. Oxford: Oxford University Press.

Wong, Hong Yu. 2019. "Emergent Dualism in the Philosophy of Mind." In *The Routledge Handbook of Emergence*, edited by Sophie Gibb, Robin Findlay Hendry, and Tom Lancaster, 179–86. London: Routledge. https://doi.org/10.4324/9781315675213-14.

8 Micro, Macro, and Meso

8.1 Emergence without the Baggage

Libertarian choice requires consciousness. We have discussed two ideas for grounding conscious experience, both of which have significant baggage. The first was interactive dualism. As we saw in Chapter 5, scientific objections to dualism are weaker than advertised. Nonetheless, the remaining philosophical questions make it an unattractive option for many. The second was strong emergence. Unfortunately, that option has nearly as much baggage as the first. It would be good, then, if there were some sort of emergence that is metaphysical—part of reality—but that had better scientific credentials than strong emergence.[1]

The goal of this chapter is to defend just such a view, what I call *moderate metaphysical emergence* (MME). It holds that higher-level properties are real and causally autonomous without requiring the mysterious "popping into existence" of strong emergence. If successful, MME would provide the metaphysical space needed for libertarian agency. Contemporary physics itself provides the evidence through three key insights: universality reveals that systems with different microstructures exhibit identical behaviors; renormalization shows why macroscopic properties depend on intermediate rather than fundamental scales; and multiscale modeling demonstrates that successful physical theories require information from multiple organizational levels. I rely primarily on the recent work of Robert Batterman, Margaret Morrison, and Mark Wilson. Their analyses reveal a striking conclusion: while many philosophers embrace fundamentalism, portraying "real physics" as exclusively bottom-up, the actual practice of physics tells a different story. Higher levels possess their own causal efficacy, not as convenient fictions but as genuine features of physical reality.

The chapter develops this argument, beginning with a deeper analysis of universality introduced in Chapter 7, showing precisely how emergent properties maintain autonomy from their micro-constituents. I then examine how the mathematical method of *renormalization* explains this

DOI: 10.4324/9781003681199-8

phenomenon, followed by an exploration of multiscale modeling—both supporting the ontological reality of non-fundamental levels in physics. The final section addresses the primary objection to MME: fundamentalists maintain that while the language of levels and emergence may be pragmatically useful, it fails to capture reality. From a God's-eye perspective, there exists only one fundamental level. Throughout the discussion, I use "level" and "scale" interchangeably,[2] referring not only to differences in size but also to variations in time and energy.

8.2 Universality and Autonomy

The argument offered here is based on physics rather than, say, chemistry or biology. While there is significant work in those fields on emergence, it is often dismissed by fundamentalists. They argue that emergence talk would be unnecessary if we knew more about nature and had the computational capacity to work from the bottom up. The truth about reality, they say, is discovered in physics. I argued against that conclusion in the two previous chapters, but there is some advantage in accepting the challenge. By all means, let's talk about physics. But we have to consider the whole of physics, not just the atomic level and below.

As noted in Chapter 7, universality and autonomy are earmarks of emergence. Universal behavior is related to what philosophers call "multiple realization," when a higher-level property is exhibited in systems with different base-level components.[3] For example, phase transitions are found among the three states of matter (gas to liquid), but also in magnets (ferromagnetic to paramagnetic) and superconductors (normal to superconducting), all of which have the same mathematical relations describing these transitions (Batterman 2019, 27–30).[4] Likewise, the relation between temperature and density is the same for entirely different fluids (Batterman 2021b, 37–39). Superconduction itself can arise from various chemical compositions.[5] Batterman and Morrison think autonomy mainly concerns understanding these patterns of universal behavior. The explanation can't be a matter of microphysics since the systems that display universality have completely different constituents (Morrison 2015, 111). Even a Laplacian demon could track the motion of every individual molecule but would still be able to explain why macroscopically dissimilar materials show the *same* critical behavior. It is the pattern—universality—which demands an explanation that transcends the micro level. Note that the standard reductionist reply to emergence does not work here, as Morrison explains:

> [The] reductionist might respond that surely it is possible *in principle* to derive macro phenomena from micro properties given the Schrodinger

> equation and the appropriate initial conditions (i.e. god could do it)... [Universality] speaks against this possibility. If we suppose that micro properties could determine macro properties in cases of emergence then we have no explanation of how universal phenomena are even possible. Because the latter originate from vastly different micro properties there is no obvious ontological or explanatory link between the micro-structure and macro behaviour... [The] mystery to be explained is how several different systems with different micro structures behave in exactly the same way; hence, because the micro structures are different in each case the explanation *cannot* be given in those terms.
>
> (Morrison 2015, 111, italics in the original)

This is a significant advance in the debate. We could know all the micro-level facts with perfect accuracy—with an exhaustive knowledge of fundamental physics and unlimited computational capacity. And yet universality would remain a mystery. The micro-facts cannot explain why systems with completely different constituents all behave the same way. They cannot explain the patterns of universality.

Here's a simple analogy. If we want to explain why a given pigment appears on some portion of a canvas, there is a physical explanation in terms of the contact force applied by the brush, the viscosity of the paint, its adherence to the canvas, coefficients of friction, etc. A separate explanation could be given for each bit of paint on the canvas, and all of these explanations collected in one document. Nothing in that compendium would explain why those bits of paint look like the Mona Lisa. The explanation for the pattern lies elsewhere.

So what does explain universality? One must show *why* microphysics plays such an attenuated role when it comes to observable phenomena. Renormalization group theory offers precisely such an explanation by revealing how macro-behavior achieves autonomy from the micro level.[6]

8.3 Renormalization

Condensed matter physics studies the properties of solids and liquids. One of its implications is that fundamental physics, like the Standard Model, contributes very little to our understanding of such systems. That by itself is not news. One can do celestial mechanics without knowing anything about quantum physics (as was done for two centuries). But the way in which condensed matter is independent of fundamental physics is different and revealing. Understanding this independence involves the mathematics of renormalization.

Before getting into the details, consider an example from Section 6.4.3 that shows in rough terms what renormalization demonstrates. Hilary

Putnam asks us to think of a wooden block with two 1-inch holes, one round and one square. A circular peg with 1-inch diameter fits in the round hole, but a square peg that has 1-inch sides will not. What explains this? It is not the composition of the board, since we would get the same outcome if the block were made of steel, plastic, or granite. The explanation lies in the macroscopic dimensions: the relative sizes of the holes and pegs determine which peg fits. Details about their composition are immaterial in this scenario.

In physics, however, identifying which properties matter is rarely so straightforward. Renormalization is a rigorous mathematical procedure for determining which properties determine a system's behavior. It effectively filters out whatever details do not influence the system's change of state over time.

Systems in condensed matter physics consist of a massive number of atoms. To get a handle on this complexity, physicists use abstract mathematical spaces such as the phase spaces mentioned earlier in Section 3.2. Another such space is composed of "Hamiltonians." A Hamiltonian represents a system's degrees of freedom and the effects of external fields. However, due to the sheer number of particles in these systems, each one contributing additional degrees of freedom, determining a realistic Hamiltonian is often impossible. Renormalization group theory is crucial for overcoming this obstacle.

Physicists and engineers often reduce the degrees of freedom when modeling a system. In statics, for example, students treat bridges as two-dimensional to analyze the distribution of force. Renormalization is a distant relative of this idea. It starts with selecting a group of Hamiltonians that includes the one thought to be the true and accurate description of the system. Again, while the correct Hamiltonian is epistemically inaccessible, one can know that it lies somewhere within a selected range.[7] Renormalization transforms this group into a new space of Hamiltonians that behaves the same way but with fewer degrees of freedom. This transformation is the mathematical heart of the procedure. The process then repeats in this new space, again eliminating and sometimes changing degrees of freedom, each time finding Hamiltonians on a larger scale. After many iterations, a single, mathematically manageable Hamiltonian emerges that exhibits the same macroscopic behavior as the inaccessible, correct Hamiltonian in the neighborhood on which the process began.

Okay, so what? It turns out that wildly heterogeneous systems can all be represented by the same simplified Hamiltonian. Prior to renormalization, the initial Hamiltonians for each of these different systems are not even in the same neighborhoods as the others. Their disparate microstructures put them in different regions of this mathematical space. Yet the Hamiltonians of many different systems flow towards the same fixed point during the renormalization process. *These systems therefore display the same macroscopic behavior, even though they are composed of entirely different constituents*—a clear example of universality. In essence, the renormalization process formally demonstrates how distinct micro level starting points can converge upon the same macro-behavior precisely because the micro-details become irrelevant at larger scales.

Let's say more about how the micro-details are "irrelevant." Here's another analogy: although Pluto's gravitational pull does affect a rocket's trajectory, it pales in comparison to Earth's gravity—so we typically ignore it. Once renormalization is complete, only the remaining degrees of freedom in the resulting Hamiltonian are truly significant. Those that matter are found at the meso level—the number of spatial dimensions, symmetry, temperature, and density (Morrison 2012, 157).[8] Small-scale, molecular interactions are insignificant when it comes to the dominant behavior of these systems. Because the small-scale details matter so little, we could swap one set of molecules with another of different chemical composition without affecting the system's overall behavior (Batterman 2021b, 40).

> [Renormalization analysis reveals] a class of macrostates of various systems at the scale of everyday objects (fluids) that are essentially decoupled or independent of their microdetails. The renormalization group explanation provides principled physical reasons (reasons grounded in the physics and mathematics of systems in the thermodynamic limit) for ignoring details about the microstructure of the constituents of the fluids.
>
> (Batterman 2011, 1037)

The "decoupling" that Batterman mentions here is what we've been calling autonomy: macro-behavior cannot be explained in terms of the micro-details. The causally relevant properties are found at a higher level. Morrison uses slightly different language:

> Part of the importance of the RG [renormalization group] is that it shows not just that we can focus on the energies or levels we are interested in, leaving out the rest, as we sometimes do in idealization and model building; it also illustrates and explains the ontological and epistemic independence between different energy levels-the defining features of emergent phenomena.
>
> (Morrison 2015, 107)

While autonomy and independence are prominent themes in the emergence literature, renormalization provides a mathematically rigorous way to understand what they mean.[9]

Let's be clear. The point is not that higher-level systems do not in any way depend on lower levels. Without a base level, there can be no emergent levels. The question is about what matters when it comes to the behavior of a system. All incompressible, 1-inch square pegs will fail to fit into 1-inch round holes regardless of their molecular makeup. What explains this pattern of behavior? The higher-level geometric properties of the peg and board. Those are the properties that matter. Molecular events are in that sense irrelevant. Renormalization tells us what matters when it comes to the systems it studies. What it shows is that the dominant behavior of those systems is driven by meso-level properties. The micro is in that sense irrelevant.

Morrison emphasizes that renormalization is not just eliminating degrees of freedom (2021, S5974), as Jessica Wilson's account of emergence would have it (Section 7.5). Nor is renormalization just a kind of averaging over microstates. RG iterations reveal a qualitative change in the causally relevant interactions. "[Successive] transformations give you a Hamiltonian for an ensemble that contains very different couplings from those that governed the initial ensemble" (Morrison 2012, 143). A "coupling" is just an interaction between particles or fields, such as a force. New kinds of interactions emerge from renormalization beyond those between molecules. These are described by macroscopic coupling constants that are causally relevant for phenomena like magnetization and heat capacity (Bishop 2024, sec. 5.2).

This is enough to show how condensed matter physics supports moderate metaphysical emergence. First, neither reduction nor weak emergence can account for universality. Complete, ideal knowledge of the micro level would still not explain why heterogeneous systems have the same macro-behavior. Renormalization does. It reveals which degrees of freedom are relevant to macro-behavior and that heterogeneous systems have those same degrees of freedom. Second, meso-scale properties are real (metaphysical). They are causally relevant for macro-behavior.

Critics are unfortunately not impressed with renormalization. They describe it as simply another example of coarse-graining—taking averages over the different parcels of Hamiltonians at each step. There is nothing intrinsically different here, they say, than things like mean kinetic energy and heat. If correct, then RG analysis at most supports weak emergence (Palacios 2022, 40).

Things are not that simple, however. It is true that coarse graining in statistical mechanics is mostly about taking averages. What is going on at the molecular scale averages out to thermodynamic properties at larger scales. But that is not how renormalization works (Bishop 2024, sec. 5.1). It proves instead that what is going on at the micro level is unimportant when it comes to the behavior of the macro level. If renormalization were just a kind of averaging, it would not have been worthy of a Nobel Prize (Kenneth Wilson in 1982).

To recap, renormalization explains the sense in which macro-scale phenomena are autonomous from micro-scale details. Through iterative transformations of Hamiltonians, renormalization reveals how systems with different microstructures exhibit the same macroscopic behavior (universality). Meso-level factors—such as density fluctuations or symmetry—play the dominant causal role, not the properties found at the lowest physical levels. Renormalization thus supports MME by showing that meso-level properties are real, causally relevant, and independent of micro-constituents.

Debates about renormalization continue apace, especially in the philosophy of science literature on effective field theories in particle physics (Palacios 2022, sec. 4) (Carroll 2022). Renormalization has become (yet another) meandering controversy within analytic philosophy. Fortunately, one can draw the same conclusion for upper-level autonomy without resorting to renormalization. Multiscale modeling in physics and engineering illustrates the causal significance of intermediate scales and the limits of purely bottom-up approaches. Batterman shows how this works in terms of hydrodynamics (2021b, chap. 3) and Brownian motion (2021b, chap. 4). Instead of those, let's consider the humble example of steel (M. Wilson 2017, chap. 5).

8.4 Multiscale Modeling

Fundamentalists believe that the behavior of observable materials, like steel, is wholly determined by their small-scale structure. At the micro level, steel is composed of lattices of molecules (Figure 8.1). At the macro level, it's a continuous material that flexes and compresses, which is what makes it so valuable in construction. At this level, Hooke's law describes how the steel will deform when a force is applied. Marble in ancient structures by comparison was strong but brittle. It fractures instead of bending and reforming.

Between the levels of macroscopic steel and the molecular lattice are several meso levels, each with its own properties and dominant behavior. Ferrite is soft and pliable, with a mostly iron crystal structure and a tiny amount of carbon. Cementite has more carbon and is hard and brittle. Pearlite contains alternating layers of ferrite and cementite on a scale of around 0.2 μm. There are different dominant behaviors at different scales. At some scales, the material will behave the same way no matter which direction a force acts on it. Other scales are much stronger in some directions compared to others. This gives each level its own *representative volume element*: tiny volumes far above the atomic scale that characterize this

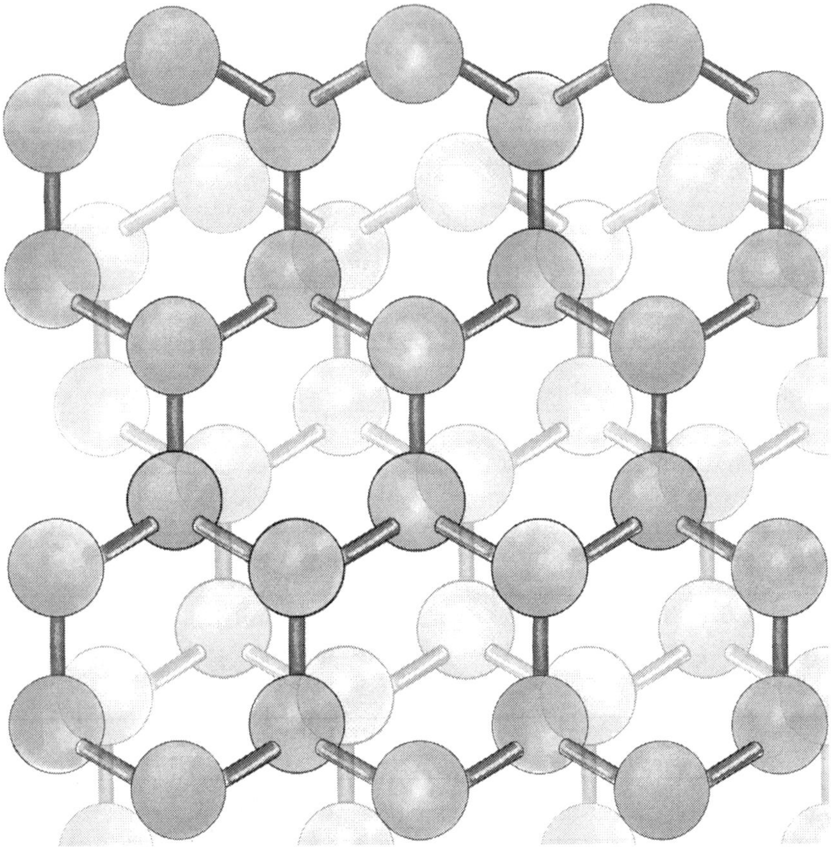

Figure 8.1 Molecular lattice.

Source: Artwork by Ashley McNutt-Greene. Used with permission.

dominant behavior. Not only do these levels have their own properties, but they also interact with one another. Let's consider one of Mark Wilson's examples.

Consider steel that is used to form railway tracks. The wheel of a train exerts a force on the rail. If the rail were perfectly elastic—an idealization—the steel would flex and then reform to its original shape. In the real world, energy at the macro level moves down through the meso levels, sometimes causing *dislocations*. From a molecular point of view, dislocations are irregularities in the lattice structure. At a meso level, they appear as squiggly lines or imperfections in the material, like those shown in Figure 8.2. Intuitively, such imperfections seem bad—something one would want to minimize. But dislocations can move under stress, acting as a buffer keeping the energy introduced by the macro-force from transmitting all the way down to the lattice. Without dislocations, the stress would go right to the molecular bonds, and the steel would fail (e.g., fracture or buckle). Dislocations and their movements shield the micro-scale structures from harm, resulting in a stronger steel beam. This is why folding and hammering by blacksmiths improves the quality of steel. The process introduces dislocations.

There is, however, a tipping point. When dislocations accumulate and can no longer move, they fail to provide this buffering effect. The steel becomes less pliable and can no longer resume its shape after flexing—one example of an effect known as "hysteresis."

Modeling strategies that work exclusively from the bottom-up or the top-down fail to capture meso-scale structures, like dislocations. This is why those methods work only in special cases. They miss the causal relevance of the meso level.

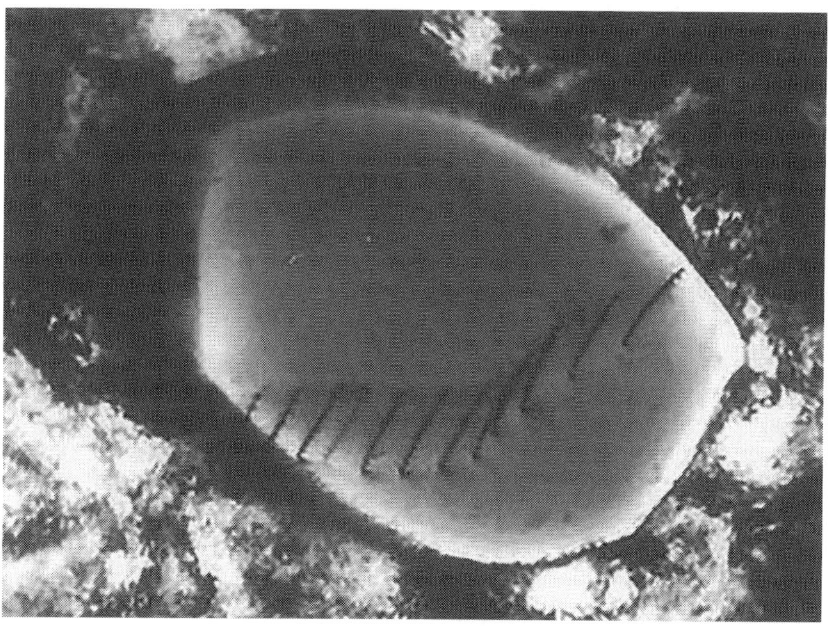

Figure 8.2 Dislocations Wikityke at the English-language Wikipedia.

Notice that the story here is almost entirely about the relation between macro-forces and meso-effects. The micro level contributes little to our understanding of, say, hysteresis, although that is one of the main properties of interest from steel beams to batteries:

> It turns out that the actual details at the level of the atomic lattice are largely irrelevant to the continuum scale behavior of the steel. This is important and reflects the *relative autonomy* of the continuum equations (the Navier–Cauchy equations and, for that matter, the equations of thermodynamics) from the detailed atomic and molecular configurations of the various systems.
>
> (Batterman 2021a, 217)

This autonomy is the common thread between renormalization and MME more generally. Macro-level theories like fluid mechanics are indifferent to the micro level but not the meso level.

The same story can be told about biological systems. Bones appear mostly homogeneous at the macro level but have distinct meso-structures at the millimeter scale (trabecular vs. cortical bone) and still more at tenths of a millimeter, micrometer, and nanometer scales (Batterman 2021b, 114–16). What matters for, say, the risk of fracture? Again, it is the dominant behavior of the meso level, not its micro-constituents.[10] The risk of fracture and healing rates depends crucially on how these meso-level layers redistribute stress. If one tries to predict fractures using only micro-scale details (e.g., the bonds in hydroxyapatite crystals) or purely macro-scale properties, the results are inaccurate. Biomedical engineers design implants and treat fractures by manipulating and supporting these meso-level structures (e.g., reinforcing trabecular bone). This direct manipulation of meso-level design features changes patient outcomes.

These insights have been incorporated into multilevel modeling, which has allowed physicists and engineers to solve the problem known as "the tyranny of scales" (Batterman 2013). Typically, the mathematical representation for any given level has ramifications for all scales below. For example, the equations in fluid mechanics treat matter as a true continuum with no gaps between discrete particles. That, of course, is not accurate. Far more structure exists at the meso and microlevels. But the same problem arises at the meso levels. Each level is represented by partial differential equations, and each entails a particular structure for matter all the way down. Every scale is "tyrannical" insofar as it dictates the structure of the material at all lower scales. The problem is that these greedy descriptions at each level are mathematically inconsistent with one another.

Multiscale modeling resolves these inconsistencies between levels by incorporating information from each.[11] (These techniques are known as *homogenization methods* (M. Wilson 2017, 214), which are related to renormalization.) Information at one level is used to correct tyranny-of-scale overreach in the others. The approach is neither bottom-up, working from a molecular base level as reductionists prefer, nor top-down, allowing the macro-level theories to dictate the structure of matter at smaller scales. Multiscale modeling takes advantage of the distinct behavior of meso levels to accomplish what more traditional methods cannot. Other examples can be found in geology (M. Wilson 2017, 21), fluid dynamics (Bishop et al. 2022, 85–87), and even cosmology (Massimi 2018).

Both multiscale modeling and renormalization demonstrate the autonomy of the macro level from its micro-constituents. This autonomy is not complete independence, as one might find in cases of strong emergence. Instead, the macro level's dependence is on the properties of the meso level, and only indirectly on the micro level. For observable materials like steel, what occurs at the smallest scales is irrelevant. The meso-properties are what matter.

Let's sum up the chapter thus far. Moderate metaphysical emergence begins with universality: systems with entirely different microstructures behave the same way at larger scales. This universal behavior cannot be explained in reductionist terms. Even an ideal knowledge of the constituents cannot explain why heterogeneous systems behave in identical ways. Universality is explained by renormalization: The macro level is autonomous of the micro level (not completely, as the analogy with Pluto illustrates; there is some causal contribution, but it is small compared to others.) Meso-level properties mediate between the extremes. The goings-on of the small-scale constituents are irrelevant. As far as a steel rail is concerned, the micro level could just as well be continuous as atomic. The properties of the rail do not depend on the micro level having a particular structure. All this is confirmed by the success of multiscale modeling, which relies on information available at various meso levels.

There is still a crucial objection that proponents of emergence must address: even if multiscale modeling works, does this reflect genuine metaphysical emergence or merely pragmatic convenience?

8.5 From Pragmatic Model to Metaphysical Reality

Let's begin with the question of what exactly "mathematical modeling" means. Do the equations describe or are they merely patterns in the data that we can exploit for our own purposes? After all, there are lots of useful mathematical models, but useful is not the same as real. Geocentric models were extremely useful. Much of machine learning today concerns finding statistical correlations in vast data sets. Whether these patterns correspond to causal relations in nature is a different question altogether.

More generally, is this talk of emergence and levels just a convenient way of discussing systems that, in reality, are just collections of atoms? Consider convection cells again (Section 7.4). We saw that macroscopic heating plates are a necessary "stability condition" for the emergence of such cells. But since the plates themselves are composed of atoms, we need not think of them in macroscopic terms (J. M. Wilson 2021, 139, n8). Both the plates and the fluids in this example can be reduced to their constituents. In the end, according to the fundamentalists, it's all just atoms. What the emergentist needs to show, says Hossenfelder, is that the plates—and stability conditions in general—cannot be described in base-level terms (2019, 92).[12] The same goes for some of the higher-level degrees of freedom in renormalization, like pressure. If they can be so described, then there is nothing intrinsically "higher level" about them. Emergence talk should not lead us to reify levels.

Carroll, again, thinks of many different scientific descriptions as referring to a single physical reality.[13] One might use the language of particle physics, while others use the language of biochemistry or neuroscience. The choice is a matter of convenience. He believes that higher-level theories are just coarse-grained descriptions of the micro-reality. For example, many molecular configurations map to one set of values for pressure, volume, and temperature. That's not surprising. At any point in time, the air molecules in the room each have a specific location and are moving in a given direction. If one of those molecules were traveling toward the ceiling instead of the floor, there would be no difference in the temperature in the room. In Carroll's view, pressure and temperature readings are coarse-grained descriptions of lower-level behavior. Even talking of particles is "useful and efficient" but not strictly speaking true (Carroll 2024, 101). All the machinery of condensed matter physics and the jargon of the special sciences is "enormously more computationally efficient than the microscopic molecular theory" (Carroll 2016, 99).[14] Nonetheless, higher-level descriptions are not strictly necessary (Carroll, n.d., loc. 1:31–32).

> If you give me the precise and complete quantum state corresponding to "a person feeling an itch," and I have the calculational abilities of Laplace's Demon, I could predict with extraordinary accuracy that the quantum state will evolve into a different state corresponding to "a person scratching themselves." No further information is needed, or allowed.
>
> (Carroll 2016, 374)

For a Laplacian demon, who has no need for computational efficiency, there is only one level. From its point of view there is no strong emergence, there is no autonomy of scales, and there is no MME.

The claim that emergence talk is wholly pragmatic and instrumental is difficult objection to disprove (as are most philosophical claims). Multiscale modeling works. Renormalization methods work. But the critics ask why any of this has metaphysical implications. They think macro- and meso-descriptions are merely phenomenological and so not philosophically important. Why think otherwise?

There are three reasons. The first is that structures at different levels interact. They cannot be alternative descriptions of one and the same physical reality if they causally affect one another. A country can be denoted by its name or by giving detailed coordinates of its borders. The former is typically more convenient than the latter, but the two cannot interact with one another since they are the same entity. In contrast, when energy cascades down through the levels in the steel example, meso-level structures play a causal role (M. Wilson 2017, 213–14, 218–20).[15] Dislocations dissipate energy and thereby shield the molecular lattice from harm. Multiscale modeling makes no sense if the macro and meso levels are simply alternative descriptions of a micro-reality.

Alan Love offers a similar defense of biological levels (2021, sec. 7.2).[16] There are criteria for when to treat a theoretical posit realistically. One is the ability to successfully intervene and redirect its behavior. Following Ian Hacking's argument for the reality of unobservable particles like electrons (1983), the ability to manipulate entities in different ways points to their reality. In short, if you can do things with them (e.g., build electron microscopes and lasers), they must be real. Love argues that the same idea applies to biological levels. Biologists manipulate structures at various levels in tissue engineering and experimental embryology to bring about different results. Contrast this with, say, sonogram "sparkles." While their appearance indicates the presence of kidney stones, they do not themselves correspond to anything anatomically (Moore 2022). They cannot be manipulated to bring about some alternative outcome. You cannot *do* things with sparkles. In contrast, Love describes how biologists causally interact with structures at different levels of organization. This successful manipulation supports the reality of those structures.[17] Going back to the steel example, the fact that dislocations can be introduced and then manipulated speaks to their reality.

The second reason is what philosophers call an "indispensability argument." Multiscale modeling is not just useful. Due to the tyranny of scales, one cannot correctly form a high-level continuum theory without information from the meso level (Batterman 2021b, 138). An exclusively bottom-up (from the molecules) approach typically fails because the properties of micro-constituents are not what determine macro-level behavior, as renormalization proves. The meso level is not dispensable.

Batterman offers a third reason: The meso-scale gets the physical ontology correct (2021b, xi, 106, 145). It "carves nature at the joints" in ways neither continuum nor atomic theories do. How so? Because of the tyranny of scales, continuum mechanics provides an inaccurate description of lower levels. Matter is gappy at the atomic scale (rather than continuous). Atomic physics, on the other hand, has limited explanatory power. As we've seen, it cannot explain universality. It is the meso level that both corrects macro-scale errors and describes heterogeneous systems. The observable world depends on meso-properties in a way that does not depend on micro-constituents that vary from system to system. This "middle-out" approach

> to many-body systems emphatically is *not* motivated only by pragmatic considerations. The mesoscale properties that connect the "fundamental" theory to the experimental data (often available only at considerably higher scales) are genuine features that are necessary for properly relating "fundamental" theories to non-fundamental theories.
>
> (Batterman 2021b, 110–11)

Meso-level properties are "objectively present in nature, quite independently of whether we recognize their exploitable advantages or not" (M. Wilson 2021, 193). Structures at those levels have their own dominant behaviors in the real world, which multiscale modeling takes advantage of. We might call it multiscale modeling, but it isn't *just* modeling. Meso-level properties are real, and their existence supports MME.

These three arguments—the causal interaction between levels, the indispensability of meso-level information, and the descriptive accuracy of the meso-scale—collectively undermine the claim that emergence is merely a pragmatic fiction. Rather, they entail that meso-level properties represent genuine features of reality that cannot be eliminated. It is not enough for the fundamentalist to point out that the higher levels depend on the lower. Emergentism grants that. The question is whether the lower levels fully determine the behavior of the higher. Renormalization and multiscale models show that they do not.

It is an article of faith among fundamentalists that derivations from higher to lower levels are still possible "in principle." With this qualifier in hand, one might even claim that elections could be predicted from knowledge of particle physics alone (Hossenfelder 2019, 90). The "in principle" response is treated as an irrefutable catch-all, sweeping aside all evidence to the contrary.

Doing so at this point simply ignores the physics discussed in this chapter.[18] Meso-level structures across multiple domains—condensed matter, fluid mechanics, engineering—consistently emerge as the driver of macroscopic behavior. This is why exhaustive knowledge of molecular lattices is inadequate to model material like steel. The behavior of what we can observe depends on causally relevant structures at intermediate levels.

> The ideal of *in principle* derivation … from more "fundamental" lower-scale details … is largely mistaken. Any examination of the actual practice of scientists interested in modeling systems at different scales will reveal nothing as simple as the kind of derivation that proponents of this ideal believe is possible…
>
> (Batterman 2021b, 45)

I have no illusions that the arguments here will forestall the appeal to "in principle" reductions in the future. The burden of proof, however, lies with those claiming that undiscovered mathematical frameworks would succeed where current ones fail. The technical obstacles identified—the tyranny of scales, the causal relevance of meso-structures, the context-dependence of emergence—are not merely computational inconveniences but reflect the mathematical structure required to accurately represent physical systems. To claim these obstacles could be overcome with greater insight and computational power is to claim that nature's apparent hierarchical organization is merely an artifact of our limited understanding. Yet advances in physics continue to reveal more, not less, organizational complexity. Fundamentalist gestures towards what they believe is possible do not solve the problem.

8.6 What Work for Moderate Metaphysical Emergence?

The primary goal of this chapter was to defend a type of metaphysical emergence that has more scientific support than strong emergence. MME fits this description. It shows that particle physicists are usually not the right people to ask about the relation between levels. Fundamental physics can tell us all sorts of interesting things about the ground floor of reality but is limited when it comes to drawing out its implications at larger scales. As renormalization shows, the dominant behavior of higher levels is due to factors other than the activity of atoms and their constituents.

This autonomy should give physicists pause before making blanket pronouncements about human psychology. Free-will skeptics do not see how to square libertarianism with physics. So what? Free will is a higher-level phenomenon. One shouldn't expect physics to have much to say about it. When physicists argue that free will is impossible, they are making a claim about human activity far above the level of physics. We can grant Carroll's argument that there's no place for libertarian freedom in the Standard Model. There is no "free-will field" in quantum field theory. Again, so what? One cannot thereby conclude that there is no libertarian freedom in the whole of reality. Such inferences ignore the autonomy of levels. We cannot foresee what higher-level phenomena will and will not unfold from atomic physics.

One still might wonder what is it that emerges at the level of human persons that allows for free will. It would presumably be the same thing that allows for consciousness, which immediately presents a problem. We have no explanation for consciousness—the so-called "hard problem" (Chalmers 1995). I suspect the solution to that problem has something to do with emergence but will not attempt to offer one here. (Nor hint that I will do so in the future, as philosophers often do.) The aim of this present book is not so lofty. I merely wish to show what is and is not in conflict with physics. Thus far, fundamentalism is doing all the heavy lifting for free-will skepticism. But fundamentalism is a philosophical commitment that is neither required nor supported by physics. MME shows there is more going on at higher-level sciences than mere coarse graining of atomic phenomena. Once that fact is recognized, claims about non-fundamental physics begin to look extravagant and unfounded. Libertarianism might be false, but not even a Laplacian demon could infer that from atomic physics alone.

So are we done? Not quite. I have argued that levels are causally related but have not said how. Doing so involves yet another thorny question, this one about downward causation. Carroll seems to believe that this is where the conflict with physics is most evident. We will take that up in the next chapter as we try to connect the remaining dots between MME and an explicitly libertarian program.

Notes

1 Contextual emergence (Section 7.4) fits this description. The view defended in this chapter would likely count as a type of contextual emergence (Bishop et al. 2022, 84).

2 The lowest levels correspond to the smallest scales. While there are some problems with this identification of levels and scale (Love 2021, 147; Robertson and Wilson 2024, 5), I ignore those here. Emergence and levels are related but strictly speaking distinct ideas. For our purposes, the former is the more important of the two.

3 The concept of multiple realization is discussed in detail in Section 6.4.3.

4 Technically scaling laws and their critical exponents (Batterman 2019).

5 For a discussion of universality in biological systems, see Ross (2025, sec. 3.3).

6 The details are technical, and a simplified description is provided here. See Batterman (2021b, sec. 2.2.3) and Williams (2023, sec. 4.3) for more.
7 Simple analogy: When I see a yellow banana, I can determine the range of frequencies of visible light that is being reflected from its surface without knowing its precise value.
8 Morrison argues that the relationship between the micro and macro levels is therefore not a matter of supervenience (2012, 164). Condensed matter macro-behavior is not determined by micro-events.

> The dependence relation required for supervenience is clearly lacking in cases of "universal" behaviour since fixing the subvenient properties in no way fixes the supervenient ones and vice versa ... The claim so often associated with supervenience—there can be no A difference without a B difference (where A properties supervene on B properties)—is irrelevant here since once the system reaches critical point and universal behaviour (A properties) is dominant, information about micro-level structure (B properties) is simply lost.
> (Morrison 2015, 112)

9 This autonomy explains a puzzling fact in the history of physics (Batterman 2021b, 139). Continuum mechanics and fluid mechanics are not based on atomic physics and yet continue to be used successfully. How did these theories survive the decisive evidence for atoms discovered in the early twentieth century? The answer is that the behavior described by these theories is not determined by atomic-level physics. Renormalization shows that the micro level was always irrelevant.
10 See Noble (2012) for detailed multi-scale interactions in hearts.
11 See, for example, COMSOL Multiphysics for a software package that specializes in this approach.
12 Some of the stability conditions discussed in Section 7.4 cannot be reduced in this fashion. One was that stable planetary orbits require three-dimensional space (Bishop et al. 2022, 29). That sort of stability condition has nothing to do with the system's constituents.
13 Carroll endorses Daniel Dennett's idea of "real patterns," which defends the use of upper-level descriptions when they have explanatory power and lead to successful predictions (Carroll 2019, 238; Dennett 1991).
14 Some critics do not accept this pragmatic justification for levels talk Potochnik doubts its heuristic value (Potochnik 2021, sec. 3.2). Others believe that it hinders a more reductive approach (Ladyman and Ross 2007, 54).
15 In what sense "causal" will be considered in Section 9.2.3.
16 In fact, a stronger case for MME can be made using biological examples (Ellis 2025; Green 2021; Green and Batterman 2021). The problem is that fundamentalists dismiss the special sciences as, again, pragmatic and instrumental. This is why I have emphasized examples from physics.
17 This manipulation argument fits nicely with James Woodward's more general account of causation, which many of the authors cited in this chapter are favorable towards (Woodward 2003; Woodward 2021). See Section 9.2.3 for more.

18 "I sometimes wish that word processors might administer a mild electric shock to authors whenever they type the phrase 'in principle,' for it's generally a signal that important issues have been lightly brushed aside" (M. Wilson 2017, 21).

Works Cited

Batterman, Robert W. 2011. "Emergence, Singularities, and Symmetry Breaking." *Foundations of Physics* 41 (6): 1031–50. https://doi.org/10.1007/s10701-010-9493-4.

Batterman, Robert W. 2013. "The Tyranny of Scales." In *The Oxford Handbook of Philosophy of Physics*, edited by Robert W. Batterman. Oxford: Oxford University Press.

Batterman, Robert W. 2019. "Universality and RG Explanations." *Perspectives on Science* 27 (1): 26–47. https://doi.org/10.1162/posc_a_00298.

Batterman, Robert W. 2021a. "Multiscale Modeling in Inactive and Active Materials." In *Levels of Organization in the Biological Sciences*, edited by Daniel S. Brooks, James DiFrisco, and William C. Wimsatt. Cambridge: MIT Press. https://doi.org/10.7551/mitpress/12389.003.0015.

Batterman, Robert W. 2021b. *The Middle Way: A Non-Fundamental Approach to Many-Body Physics*. Oxford: Oxford University Press.

Bishop, Robert C. 2024. *The Physics of Emergence*. 2nd ed. San Rafael: IOP Publishing. https://doi.org/10.1088/978-0-7503-6367-9.

Bishop, Robert C., Michael Silberstein, and Mark Pexton. 2022. *Emergence in Context: A Treatise in Twenty-First Century Natural Philosophy*. Oxford: Oxford University Press.

Carroll, Sean. 2016. *The Big Picture: On the Origins of Life, Meaning, and the Universe Itself*. New York: Dutton.

Carroll, Sean. 2019. *Something Deeply Hidden: Quantum Worlds and the Emergence of Spacetime*. New York: Dutton.

Carroll, Sean. 2022. "The Quantum Field Theory on Which the Everyday World Supervenes." In *Levels of Reality in Science and Philosophy*, edited by Stavros Ioannidis, Gal Vishne, Meir Hemmo, and Orly Shenker. Jerusalem Studies in Philosophy and History of Science. Cham: Springer International Publishing. https://doi.org/10.1007/978-3-030-99425-9_3.

Carroll, Sean. 2024. *Quanta and Fields: The Biggest Ideas in the Universe*. New York: Dutton.

Carroll, Sean. n.d. *Solo: Emergence and Layers of Reality*. Episode 295. Mindscape. 1:35. Accessed November 13, 2024. www.preposterousuniverse.com/podcast/2024/11/11/295-solo-emergence-and-layers-of-reality/.

Chalmers, David. 1995. "Facing up to the Problem of Consciousness." *Journal of Consciousness Studies* 2: 200–219.

Dennett, Daniel. 1991. "Real Patterns." *The Journal of Philosophy* 88 (1): 27. https://doi.org/10.2307/2027085.

Ellis, George F. R. 2025. "Biological Emergence: A Key Exemplar of the Open Systems View." In *Open Systems: Physics, Metaphysics, and Methodology*, edited by Michael E. Cuffaro and Stephan Hartmann. Oxford: Oxford University Press.

Green, Sara. 2021. "Cancer beyond Genetics: On the Practical Implications of Downward Causation." In *Levels of Organization in the Biological Sciences*, edited by Daniel S. Brooks, James DiFrisco, and William C. Wimsatt. Cambridge: MIT Press. https://doi.org/10.7551/mitpress/12389.003.0014.

Green, Sara, and Robert Batterman. 2021. "Making Sense of Top-Down Causation: Universality and Functional Equivalence in Physics and Biology." In *Top-down Causation and Emergence*, edited by Jan Voosholz and Markus Gabriel. Synthese Library 439. Berlin: Springer.

Hacking, Ian. 1983. *Representing and Intervening: Introductory Topics in the Philosophy of Natural Science*. Cambridge: Cambridge University Press.

Hossenfelder, Sabine. 2019. "The Case for Strong Emergence." In *What Is Fundamental?*, edited by Anthony Aguirre, Brendan Foster, and Zeeya Merali. The Frontiers Collection. Cham: Springer International Publishing. https://doi.org/10.1007/978-3-030-11301-8_9.

Ladyman, James, and Don Ross. 2007. *Every Thing Must Go: Metaphysics Naturalized*. Oxford: Oxford University Press.

Love, Alan C. 2021. "Manipulating Levels of Organization." In *Levels of Organization in the Biological Sciences*, edited by Daniel S. Brooks, James DiFrisco, and William C. Wimsatt. Cambridge: MIT Press. https://doi.org/10.7551/mitpress/12389.003.0011.

Massimi, Michela. 2018. "Three Problems about Multi-Scale Modelling in Cosmology." *Studies in History and Philosophy of Science Part B: Studies in History and Philosophy of Modern Physics* 64 (November): 26–38. https://doi.org/10.1016/j.shpsb.2018.04.002.

Moore, John. 2022. "Distinguishing Scientific Results from Experimental Artifacts." Phys.Org, August 5. https://phys.org/news/2022-08-distinguishing-scientific-results-experimental-artifacts.html.

Morrison, Margaret. 2012. "Emergent Physics and Micro-Ontology." *Philosophy of Science* 79 (1): 141–66.

Morrison, Margaret. 2015. "Why Is More Different?" In *Why More Is Different*, edited by Brigitte Falkenburg and Margaret Morrison. The Frontiers Collection. Berlin: Springer. https://doi.org/10.1007/978-3-662-43911-1_6.

Morrison, Margaret. 2021. "Turbulence, Emergence and Multi-Scale Modelling." *Synthese* 198 (S24): 5963–85. https://doi.org/10.1007/s11229-018-1825-5.

Noble, Denis. 2012. "A Theory of Biological Relativity: No Privileged Level of Causation." *Interface Focus* 2 (1): 55–64. https://doi.org/10.1098/rsfs.2011.0067.

Palacios, Patricia. 2022. *Emergence and Reduction in Physics*. Cambridge: Cambridge University Press. https://doi.org/10.1017/9781108901017.

Potochnik, Angela. 2021. "Our World Isn't Organized into Levels." In *Levels of Organization in the Biological Sciences*, edited by Daniel S. Brooks, James DiFrisco, and William C. Wimsatt. Cambridge: MIT Press. https://doi.org/10.7551/mitpress/12389.003.0007.

Robertson, Katie, and Alastair Wilson. 2024. "Introduction." In *Levels of Explanation*, edited by Katie Robertson and Alastair Wilson. Oxford: Oxford University Press.

Ross, Lauren N. 2025. *Explanation in Biology*. Cambridge: Cambridge University Press. https://doi.org/10.1017/9781009300940.

Williams, Porter. 2023. *Philosophy of Particle Physics*. Cambridge: Cambridge University Press. https://doi.org/10.1017/9781009205382.

Wilson, Jessica M. 2021. *Metaphysical Emergence*. Oxford: Oxford University Press.

Wilson, Mark. 2017. *Physics Avoidance: Essays in Conceptual Strategy*. Oxford: Oxford University Press.

Wilson, Mark. 2021. *Imitation of Rigor: An Alternative History of Analytic Philosophy*. Oxford: Oxford University Press.

Woodward, James. 2003. *Making Things Happen: A Theory of Causal Explanation*. Oxford Studies in Philosophy of Science. Oxford: Oxford University Press.

Woodward, James. 2021. "Downward Causation and Levels." In *Levels of Organization in the Biological Sciences*, edited by Daniel S. Brooks, James DiFrisco, and William C. Wimsatt. Cambridge: MIT Press. https://doi.org/10.7551/mitpress/12389.003.0013.

9 Compatibilist Libertarianism

9.1 Is There Such a Thing?

On its face, Christian List's "compatibilist libertarianism" is an oxymoron. Traditionally, libertarianism rejects determinism while compatibilism accepts it, making them seem mutually exclusive. List attempts to reconcile these opposing views, and while I believe his approach is largely correct, his arguments have failed to gain significant traction in mainstream philosophy.[1] There is no mention of compatibilist libertarianism, for example, in the *Stanford Encyclopedia of Philosophy* entry on free will (O'Connor and Franklin 2021). Fortunately, the work done thus far in this book contains what compatibilist libertarianism lacks. The goal of this chapter is to build a positive case for libertarian freedom in light of contemporary physics, and directly address the free-will skepticism introduced in Chapter 1.

List presents three necessary (and jointly sufficient) conditions for free will (List 2019b, 22–27; List 2022, 172). Not every libertarian will agree. The three conditions are introduced here with a fuller discussion to follow.

A. The capacity to act intentionally.
   Intentions here are understood as being goal directed.
B. The capacity to choose between alternative possibilities.
   This is AP from Section 2.3.1. "A person has free will only if he could have done otherwise" (Timpe 2012, 70), if he has the ability to do other than what he actually does. Recall that AP takes the better-known Principle of Alternative Possibilities and removes any implications about moral responsibility.
C. The capacity to control one's actions.
   This is closely linked to (A) but includes the idea that mental causation is partially responsible for an agent's actions.

DOI: 10.4324/9781003681199-9

It should be clear why these conditions lead to libertarianism. Free agents can act to bring about their own goals (A), and whatever act they choose to bring about (C), they could have done otherwise (B). List is a compatibilist because he holds that each of these three conditions can be met even if physical determinism is true. That's the surprising part. The main idea is that the properties necessary for libertarian freedom emerge at higher levels of reality, which are not (wholly) determined by the fundamental physical level. While List's conclusion is correct, his framework lacks certain elements that can be found in moderate metaphysical emergence (MME).

This chapter will begin by examining List's three conditions, analyzing both skeptical challenges and his responses. For each condition, I will then show how MME provides additional resources that strengthen the case for compatibilist libertarianism. One important idea that has been left until the end is an account of downward causation. How is it that emergent consciousness can bring about bodily actions? I draw on James Woodward's difference-making view of causation to answer. Finally, I consider how emergence not only constrains the behavior of lower levels but also enables novel capacities.

9.2 Three Conditions

9.2.1 Intentional Action

List's first necessary condition for free will is goal-directed intentionality. Agents have internal representations about their environments and themselves. They have motivations and desires. And they typically have the ability to pursue those desires, at least in some limited form, which leads to goals and plans for fulfilling them. Non-living beings have none of these.[2] We don't need to know at what point agency in living beings begins (all animals? animals with particular traits?) to know that humans meet the criteria. Intentional action is highly intuitive. Why would one question it?

9.2.1.1 Skepticism about Intentions

The standard reductionist reply is that there is nothing special about intentions or agency (Caruso, List, and Clark 2020, 62). Goal-directed acts can be reduced to complex arrays of physical, non-intentional events. This must be possible (in principle!) because reality is ultimately physical and non-intentional. Living and non-living beings alike are made of atoms. Atoms and their constituents are not goal directed. In Aristotelian terms, there are no final causes, only efficient ones.

That said, reductionists agree that acting as if persons and animals have intentions is quite useful. Daniel Dennett calls this taking the "intentional stance" (1998). Thinking of some entities as having intentions helps us to predict their behavior. Of course, no one thinks that we can dispense with intentional language. Yet the utility of intentional language does not entail that intentions are real. For example, while we might attribute feelings of love to the wagging of our dog's tail (and so predict that it won't attack us), this doesn't entail that the dog really loves us. It may mean simply that tail-wagging behavior evolved as wolves entered more human environments. Attributing intentions to dogs may be useful but not reflective of the inner canine thought-life.

9.2.1.2 *List's Defense*

List has two replies. The first is what philosophers call an "indispensability argument":

> [The] notion of intentional agency is both practically and scientifically indispensable because there are many phenomena that we would not be able to make sense of without understanding humans and other complex animals as intentional agents.
>
> (List 2019b, 25)

Both understanding and predicting human behavior require the idea of intentionality.

Before considering why List takes this to be the case, let's discuss indispensability arguments more generally. Their popularity traces back to Harvard philosophers W. V. Quine (1948) and Hilary Putnam (2012). Such arguments begin with some domain, such as science or mathematics. The next step is to show that this domain depends in part on the existence of certain entities. Finally, one should be ontologically committed to any entity that is indispensable to domains that we take to be true, such as our best scientific theories. For example, Quine argued that one cannot do modern physics without sets; therefore, sets exist. We are justified in believing in the existence of whatever is indispensable to good science.

A related version of the argument says that we should be committed to those entities that are explanatorily indispensable, i.e., explanations in the domain depend on such posits (Colyvan 2024, sec. 5). This is the type of argument List uses:

> To determine whether a system is an intentional agent, we must ask whether the system is *best* explained in intentional terms. For a system

> to pass this test, it must be the case that taking an intentional stance towards the system is not merely possible, but necessary or indispensable for certain explanatory purposes.
>
> (List 2019b, 57)

Dennett's intentional stance may be useful with certain machines (e.g., the garage door thinks there is an obstruction), plants (e.g., the crocuses think it's spring), or animals (e.g., the tail-wagging example). However, List argues that intentionality is indispensable for explaining and predicting the behavior of higher animals. In humans, the social sciences presuppose that humans have intentions, beliefs, and goals (List 2019b, 61). Scientific realists should take these theories to be at least approximately true (List 2019b, 74–76).

Finally, he argues that intentionality cannot be reduced to neuroscience (List 2019b, 67–68). Beliefs and desires inherently possess "aboutness": they are directed toward things beyond the agent.[3] For example, one might believe that a solar eclipse will occur in 2030 and desire to witness it. Both the belief and the desire are about the eclipse, and they are connected by their shared content. In contrast, brain states are not about anything. They lack semantic content. Neurons change state due to electrochemical causes, not because of reasons or content. "Aboutness" and the semantic relationships between beliefs and desires cannot be reduced to non-intentional brain states or mere causal processes.

List's second reply is a multiple realization argument (2019b, 69–74). A single mental state (e.g., believing there is snow on my driveway) can be instantiated in different brains (mine, my wife's, and my neighbor who wishes I would shovel my driveway as often as he does his own). That mental state cannot be reduced to a specific brain state since many different brain configurations produce the same mental state (Bickle 2020). Even if there were a brain-level explanation for why my wife has a given mental state, that explanation would not apply to others whose brains are configured differently. Multiple realizability thereby undercuts the reductionist claim about the explanatory priority of lower levels. A list of all the different biological systems that exhibit the same higher-level property does not explain *why* those systems have that property.

9.2.1.3 Objections and Rejoinders

The two main objections to List's first criterion will sound familiar at this point in the story. First, intentional agency and other psychological notions are only indispensable as a practical matter. They are not indispensable in principle. The critic might even grant that intentional language will never be cashed out in non-intentional terms, but that merely reflects our inability to translate one imprecise discourse into another. Philosophers might worry about such things. Scientists do not.

Though List might object, his framework cannot block this kind of move. Here's where MME can strengthen his position. Section 8.5 demonstrated the inherent limitations of appealing to "in principle" reductions. Reductionism would carry more weight if physics itself followed a clean hierarchical pattern—from condensed matter physics down to the Standard Model—then extending that logic to reduce psychology to neuroscience. Yet, physics itself resists this reductionist narrative, as it presents genuine cases of emergence. MME is metaphysical in the sense that it is real and not merely useful. Chapter 8 established that meso-level structures causally interact with those at other scales. Universality reveals that the observable world depends far more heavily on meso-properties than on its microscopic building blocks. The meso level isn't some temporary conceptual scaffolding that future science will eventually replace with fundamental physics. Even with ideal physics—complete with unlimited computational resources—emergent properties will remain genuinely indispensable. Physics cannot discover all its truths by starting from the bottom and working upward, for the simple reason that not all physical truths flow upward from the fundamental level. This bottom-up picture fundamentally misrepresents the nature of physical reality.

At the very least, the case for MME gives us reason to be skeptical of in-principle reductions. If one could fully reduce the meso levels in physics to the micro level, then promissory notes about the reducibility of intentionality and the like might be plausible. Let those physicists casting doubt on free will complete the reductions in their own field, and then we can talk about higher-level phenomena. Until then, there are reasons to think intentionality is indispensable in understanding human behavior. Given the evidence for MME, the burden of proof is on the fundamentalist to show otherwise.

The second objection involves multiple realizability. In fact, it's not so much of an objection as a dismissal. List's talk of mental states and brain states concerns the messy world of biology and psychology. Critics think mental properties specified in folk-psychological terms like "belief" and "desire" are vague. Even some social scientists push back against the need for "undetermined agency" as a theoretical posit (Caruso, List, and Clark 2020, 74–75). Physicists have little to say about multiple realizability, believing it irrelevant to their concerns.

Except that it's not. While physicists do not refer to multiple realizations per se, they do talk about universality. Universality, recall, refers to systems with entirely different microstructures behaving the same way at larger scales—a pattern that cannot be explained by microphysics alone. We see this again in phase transitions (Section 8.2). Chemistry students are familiar with the three states of matter, but phase transitions also occur in magnetic materials, superconductors, and a host of other phenomena. The same equations apply to transitions among highly heterogeneous materials. Phase transitions are "multiply realized" in the many different systems in which they occur. This universal behavior cannot be explained simply in terms of microphysics since the composition of each system is very different from the others. Universality therefore resists reductive explanations. What one wants is an explanation for why diverse systems all behave the same way. Unlike mental states, universality cannot be dismissed as one more vagary of behavioral science.

Universality is a type of multiple realization found in the relatively simple world of physics. The examples there are concrete, and reductive dismissals are more strained. That universality is found in physics makes the case for multiple realizations at higher levels plausible. It provides support for both MME and List's first condition.

Universality in physics offers a concrete example of multiple realizations within a domain that skeptics cannot easily dismiss. Unlike mental states, which critics often characterize as vague, universality manifests in physical phenomena where diverse systems exhibit identical mathematical behavior despite different microscopic compositions. The empirical reality of multiple realizations in physics strengthens the plausibility of similar patterns at higher levels of organization. By demonstrating that even basic physical systems resist reduction to their microscopic details, universality provides compelling support for both MME and List's first condition of intentional agency.

While establishing the reality of intentional agency is crucial, the libertarian conception of free will hinges upon the agent's capacity to choose otherwise. This brings us to List's second condition: the requirement that agents possess genuine alternative possibilities.

9.2.2 *Alternative Possibilities*

According to leeway libertarians (Section 2.3.1), alternative possibilities (AP) are essential to free will, where AP is List's criterion (B). Even many sourcehood libertarians agree that AP is a necessary condition.[4]

9.2.2.1 Skepticism about AP

There are two main ways to attack AP, both of which have been discussed in previous chapters. The first is physical determinism (Section 3.1): Given the laws of nature and the initial conditions, there is only one physically possible outcome at any moment. Hence, there are no real alternatives. This is why most take determinism to be incompatible with libertarianism.

Even if quantum mechanics allows for some indeterministic events, nature is sufficiently deterministic at the classical level to render AP impossible.

The second objection is eternalism and the block universe. Special relativity entails that the future exists and does not change (Sections 4.1–4.3). There is no AP-branching from a God's-eye perspective.

9.2.2.2 *List's Defense*

Starting with the second objection, List does not discuss relativity or eternalism. The implications of a block universe for free will are less clear than those of physical determinism, so that isn't surprising. For my part, I believe the leeway libertarian needs to offer a reply.

As for physical determinism, List notes that while it is still an open question in physics, the answer isn't important.

> Physical indeterminism is neither necessary nor even sufficient for alternative possibilities in agency. Alternative possibilities, I suggest, do require a form of indeterminism, but what they require is indeterminism *at the agential or psychological level, not at the physical one.*
> (List 2019b, 80–81)

Even if determinism is true at the level of physics, that is not the end of the story. Determinism at the base level does not entail determinism at all levels. Physical determinism would only be a problem if there were only one level, as the reductive physicalist claims.

Let's spell this out in terms of the Consequence Argument (Section 2.2.3). The argument says that if determinism were true, then our actions would be the inevitable consequences of the laws of nature and events in the remote past over which we had no control. Since we have no power over the past or the laws of nature, and these factors completely determine our future actions under determinism, we have no control or free will over our present acts.

This argument, as List observes (2019a, 256), relies on laws of fundamental physics, specifically those described by general relativity and quantum field theory. These laws, however, say nothing about beliefs, intentions, or choices, which are emergent properties arising long after the Big Bang. The argument's validity hinges on the assumption that higher-level, emergent properties can be reduced to fundamental physical laws. This hidden reductionist premise is critical. The Consequence Argument assumes that fundamental physics is responsible for all aspects of reality. If instead we recognize the reality of emergent properties at different levels, then determinism at the fundamental physics level entails nothing about agent choices or abilities.

List's counterargument employs the concept of coarse-graining developed earlier (Sections 1.3.4, 8.3, and 8.5). Fundamentalists interpret all higher-level theories as coarse-grained descriptions, i.e., pragmatic approximations that sacrifice precision for computational tractability. But List argues that this interpretation misunderstands what occurs when we shift between levels of description. The transition from micro to macro reveals genuinely novel properties, not merely blurred details of an underlying reality.

Coarse-graining begins with the idea that for any given macro-state, there are many compatible micro-states. Think phase spaces again (Section 3.2). There are lots of ways a system can evolve from the state points within any region of a phase space. In the case of chaotic systems, nearby states can take radically different trajectories even though the system is deterministic. This is the famous "butterfly effect." From a macro-perspective, a chaotic system behaves randomly (as the name implies). There is no predictable fact of the matter in macro terms about the precise future state of the system. Randomness is found at one level, determinism at another. Chaos is an extreme example, but there are many cases where a system's "micro- and macro-dynamics do not mesh," as Jeremy Butterfield puts it (List 2019b, 97). Micro-determinism does not entail macro-determinism, and so the Consequence Argument is invalid.

9.2.2.3 *Objections and Rejoinders*

There is something odd about this exchange insofar as both List and his critics make use of coarse-graining (1.3.4). This is due in part to some looseness regarding "randomness" and "indeterminism." Carroll and Hossenfelder agree that coarse-grained theories—all those above the level of fundamental physics—often involve randomness, but randomness does not entail indeterminism. Randomness, they say, is epistemic. Recall the dice example from Section 3.4. Dice introduce an element of chance in games. Our inability to know how the dice will land makes the game more interesting. Nonetheless, the relevant laws of mechanics are deterministic. There is a fact of the matter about how the dice will land given their initial conditions. There is no randomness in the system from the point of view of a Laplacian demon.

Fundamentalists take higher-level, coarse-grained descriptions as a useful shorthand for physical reality, which is determined by microphysics. They say that if the underlying physics is deterministic, then macroscopic randomness is merely epistemic and does not provide evidence for indeterminism. As some philosophers put it, epistemic randomness falls short of ontological indeterminism. When it comes to AP, just because we cannot perfectly predict the behavior of a person tells us nothing about whether that person's choices are fully determined.

List anticipates this objection. His reply relies again on indispensability plus scientific realism about the behavioral sciences (List 2019b, 98, 102). Taking a science realistically means seeing its mature theories as at least approximately true and its theoretical terms as referring to actual entities. List argues that AP is presupposed in explanations of human behavior. I believe that coffee will help me be more productive in the morning. I have the desire to be more productive, which leads to an intention to reach for my mug. More precisely, that intention is to reach for the mug rather than muddle through as best I can. The explanation for why I reach for the mug only makes sense in light of AP. Insofar as the behavioral sciences include the choices of agents (or groups of agents) from among alternative possibilities, AP should be treated realistically. Prospect theory, for example, which was the basis of Daniel Kahneman's Nobel Prize in economics, focuses on how individuals perceive and evaluate alternative outcomes (Kahneman and Tversky 1979). The theory makes no sense without the idea of AP.

This leaves the debate at an impasse. List treats the level at which mental states are found realistically. The free-will skeptic treats it instrumentally. Mental terms and the "intentional stance" are useful, they say, but physics alone tells us what exists. The critics take List's argument to support a conditional: *If* there are levels and agents reside at a higher level than physics, then physical determinism does not entail agential determinism. But then they deny the antecedent. There is no reason to take autonomous levels realistically.

To break the impasse, one must undermine reductionist intuitions, which are largely shaped by exceptional cases. The (partial) reduction of thermodynamics to statistical mechanics has been particularly influential. While that was a startling achievement, it applies only to special cases (gases at equilibrium) and requires more than one idealization (an infinite number of particles and volume). Reductionism as a research program has failed, which is why even fundamentalists are willing to use the language of (weak) emergence.

The next step is to show that weak emergence is not sufficient. The fundamentalist is perfectly content to say that we don't yet know enough to reduce all scientific knowledge to physics. What List needs is a robust case for metaphysical emergence, the view that higher-level properties, systems, and phenomena are real (not just convenient redescriptions). The sciences above the level of physics are not merely coarse-graining a world that is nothing but a collection of fundamental particles. Recall multiscale modeling and renormalization, which are likewise not amenable to the sorts of reductive procedures found in statistical mechanics.

Let's briefly reiterate the argument. For a homogenous ideal gas, coarse-graining is effective: by averaging microscopic variables such as position and velocity, one can derive macroscopic thermodynamic properties. However, renormalization and multiscale modeling operate on fundamentally different principles. These approaches apply to systems that are neither homogeneous nor composed of ideal gases, where simple averaging fails. In these complex systems, higher levels of description are not merely coarse-grained averages of lower-level phenomena. Instead, sophisticated techniques like homogenization have been developed to extract information across multiple scales, addressing the "tyranny of scales" problem discussed in Chapter 8.[5] Renormalization provides perhaps the clearest example of higher-level autonomy—where events at higher levels are not wholly determined by those at lower levels.[6] Similarly, multiscale modeling demonstrates the ontological significance of meso-level properties, providing evidence for metaphysical emergence.

Fundamentalists treat higher levels instrumentally. The language of (weak) emergence is useful, they say, but doesn't refer to anything real. While List has difficulty blocking this move, the case for MME provides the resources to do so. The realist about levels can point to successful scientific practice across multiple disciplines. The instrumentalist must explain away this success while maintaining that only one level is "really real."

We can now return to AP. Compatibilist libertarians are willing to grant that physics is deterministic. Nonetheless, the autonomy of higher levels shows that deterministic laws within physics would not guarantee determinism at the agential level. Even meso- and macro-events can fail to meet the conditions for determinism discussed in Section 3.2. More precisely, the mathematical proofs that apply to deterministic systems (e.g., those modeled by ordinary differential equations) do not apply to much of the rest of physics, and whether they do so is context-dependent (Bishop and Beim Graben 2016; Bishop, Silberstein, and Pexton 2022, 287). Believing that determinism at the lowest level entails determinism at all levels is a reductionist fiction that is not supported by science. Unless a stronger case can be made, AP remains viable.

Before moving on, there was a second objection to AP that List says nothing about. If the standard interpretation of special relativity is correct, there are no forking paths in the future. The future is fixed in a block universe, leaving no room for AP.

Chapter 4 provided some avenues for the libertarian to reply, making room for a dynamic view of time and an open future in light of relativity. Some involve interpretations of special relativity that do not entail a block universe. Others show how general relativity allows for options regarding time that special relativity does not. While the physics of spacetime provides obstacles to AP, those obstacles are not insurmountable.

Granting, then, both intentionality and the possibility of choosing otherwise, free will further requires that our intentions can actually bring about our actions. Let's consider the case for mental causation.

9.2.3 Mental Causation

List's third condition requires an agent's mental states to be partially responsible for some actions.

> [For] an action to be due to our own free will, it must be caused by our intentions. It must be true that we did the action *because we intended to do it*. It is not enough for the action to be purely the result of some nonintentional (and subconscious) process over which we have no control.
>
> (List 2019b, 25)

My desire for more coffee and my intention to drink it are partly responsible for the motion of my hand.[7]

9.2.3.1 Skepticism about Mental Causation

There are two main objections to mental causation, one that has already been dealt with extensively and another that will require more work. The first is the causal closure of physics: every physical effect that has a cause has a sufficient physical cause. In Chapter 5, I showed that while closure is one of the pillars of physicalism, physics neither entails nor requires it. Closure is little more than a metaphysical posit that one need not accept.

The second objection was mentioned briefly (Section 7.3.3) but put off until now. Mental causation typically involves a kind of downward (a.k.a. top-down) causation, from higher to lower levels. Skepticism about such ideas is one of Carroll's main concerns when it comes to free will. "There is no reason to think that anything like downward causation really happens in the world, at least not down to the level of particles and forces" (Carroll 2016b).

Jaegwon Kim's well-known "causal exclusion argument" aims to support the same conclusion (Kim 2003). The argument starts with a common view about the relation between mental states and brain states (Figure 9.1). Causation at the level of psychology tracks the effect of, say, my desire for more light in the room (M1) on my intention to turn on a lamp (M2). Neurophysiology, on the other hand, studies the underlying causal relation between brain states (B1, B2, ...) and the dependence of the mental on the physical (vertical arrows). The question is: What does the causal arrow between M1 and M2 represent? Even if the precise nature of

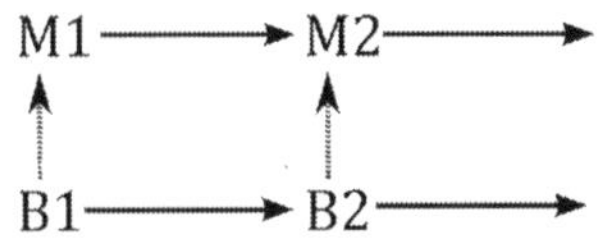

Figure 9.1 Causal exclusion.

the dependence of mental states on brain states remains mysterious, Kim takes it that once B2 is instantiated, it is sufficient for M2. (This sufficiency begs the question against substance dualism, but let's ignore that for now.) If so, what causal work is left for M1? B2 is responsible for M2, and B1 is causally sufficient for B2. Causes at the level of mental states appear to be redundant. All the causal work is done at the next level down, leaving mental states as mere epiphenomena. Folk psychological talk about desires causing intentions and the like therefore does not refer to a causal relation in nature.

If this wasn't bad enough, the same story can be told for neurophysiological causes and the electrochemical processes on which *they* depend. Considering the more fundamental causal chain one level down, causation between brain states is redundant. This sort of "causal drainage" then proceeds all the way down to the lowest level, just as fundamentalism claims. "The causal exclusion argument combined with effective field theory [in physics] is the main reason why physicists believe that reductionism is correct" (Hossenfelder 2019, 91).

For our purposes, the causal exclusion argument purports to show that there is no causal work left for mental states to do, either on their own level or in a downward sense. It might seem as if M2 influences some B3 that starts the causal chain involving the motion of my hand, but the same problem occurs. B2 is sufficient for B3. M2 has nothing to contribute. There is nothing left for downward mental causation to do.

9.2.3.2 Reply: Difference-Making Causation

Kim's causal exclusion argument has spawned a cottage industry in the last twenty-five years.[8] While List critiques the argument (2019b, 124–30), one solution lies in how we understand causation. Kim assumes a traditional (anti-Humean) productive view whereby causes bring about their effects, like when a flame causes a kettle to boil (2007, 236). An increasingly popular alternative is James Woodward's difference-making view of causation (2003). Despite not being addressed in Chapter 8, both Batterman (2021, 123–26) and Wilson (2017, 183, 252) endorse it, along with List (2019b, 131–38). Difference-making sidesteps Kim's problem by rejecting the assumption that only one cause at the lowest level can do the "causal work."

Let's start with the familiar example of controlled studies. There is an experimental group that is exposed to some chemical agent. There is a control group that is as similar as possible to the experimental group except that it is not exposed to the chemical. In the case of, say, a potential carcinogen, laboratory mice are used to see if the chemical produces a higher rate of cancer in the experimental group. If so, the study supports the claim that the chemical agent causes cancer. Call the agent *C* for cause and the higher rate of cancer *E* for effect.

Causes involve three (jointly sufficient) conditions on Woodward's account. The first is counterfactual dependence. Causal claims are evaluated based on "what-if-things-had-been-different" scenarios. If some other group of mice were exposed to *C*, then *E* would occur in that group. If the actual group had not been exposed to *C*, then *E* would not have occurred. But counterfactual dependence is not sufficient for causation. Whether the snow in my driveway melts is counterfactually dependent on what the outdoor thermometer on my garage reads, but the thermometer does not cause the snow to melt.

The second is the ability to manipulate the cause to bring about different effects. If we think of *C* and *E* not as events but as variables measuring the properties of some system, then experimental changes to *C* should produce corresponding changes to *E*. In our example, greater exposure to the chemical agent should produce a higher rate of cancer in the experimental group, less exposure produces a lower rate.

The third is invariance. The intervention used to manipulate *E* must work across a range of circumstances. Take the snow example again. If I put a candle under the thermometer, the gauge changes, but there is no change in snow. When the air temperature rises, the snow consistently melts, regardless of whatever other everyday events are happening nearby. This shows that the relation between the behavior of the thermometer and the snow melting is merely a correlation. The relation between the rising air temperature and the snow is causal.

Difference-making resolves causal exclusion by evaluating causal relationships independently at each relevant explanatory level (Woodward 2008, 234–35). If a lower-level relation satisfies the criteria for difference-making, it qualifies as causal. Similarly, if a higher-level relation meets these criteria, it also qualifies as causal—even if this higher-level relation has some form of dependence on the lower-level one (as occurs in cases of emergence, i.e., a base level has to exist for properties to emerge from it). Satisfying the difference-making criteria is sufficient for causation *at that level*, regardless of lower-level processes. Thus, explanatory levels do not compete with one another, as Figure 9.1 might suggest. The same effect can legitimately have multiple causes operating simultaneously at different levels, provided each cause identifies a difference relevant to the occurrence of that effect. The causal exclusion problem arises only from adopting a productive conception of causation.

Applying this to mental causation, mental properties have difference-making effects (List 2019b, 138). For example, a mental state like "believing I forgot to pay a bill" can begin a causal sequence that ends with me writing a check. One can intervene on that belief through persuasion or suggestion (e.g., my wife says she already paid the bill), which in turn changes the behavior (not writing the check). There is a counterfactual dependence, and the relation holds across a range of circumstances (where I happen to be in the house, what time of day it is, etc.). The belief therefore meets the criteria for a difference-making cause.

Although it wasn't explicit earlier, difference-making causation is a part of MME. This helps to answer a question that might have arisen in some readers' minds: Why isn't MME just a type of strong emergence? While there is overlap between the two, there are also some key differences. The first is that compared to strong emergence, MME is quite common. Examples can be found throughout the sciences—bone structure, the heart, etc.[9]—even though I have focused on those in physics. The second concerns the nature of causation. Strong emergentists typically hold a metaphysically thick, productive view of causation. O'Connor, for example, defends a type of agent causation (Section 2.4.3) in which persons bring about effects within a strong emergence framework (2000, 68). Strong emergence is another way to provide a metaphysical foundation for free will, alongside dualism and MME. As we've seen, however, the case for strong emergence is built on a small number of examples, each of which is problematic in some way or other (Section 7.3). But, since we don't need strong emergence to show that physics is compatible with libertarian freedom, I leave it to others if they wish to make that argument.

We've seen how difference-making causation resolves causal exclusion. There is a way to make sense of causes at the level of mental states, even if their existence depends on lower levels. ("Depends," again, in that a base level must exist for higher levels to emerge from it.) But that is not the whole problem. There is still the question of how those mental states bring about lower-level changes. While Carroll treats mental causes and effects as handy, macro-level descriptions, he rejects downward causation (Carroll 2016a, 375). Particles and fields, he says, are only acted upon locally—by the particles and fields around them. There is no sense in which they are acted upon from higher levels.

Fortunately, this final step regarding mental causation has not gone unnoticed. The reply will show not only how downward causation is possible but also allow us to fill out one more aspect of MME.

9.2.3.3 *Difference-Making and Downward Causation*

As we saw in Chapter 8, multiscale modeling exploits information at different levels to solve the tyranny of scales problem. But are the structures and properties at different levels *causally* related? I claimed that they are but never explained the notion of causation being used. Now we have the tools to do so.

The apparent paradox of downward causation often arises from a productive conception of causes as physical forces that push things around, a perspective exemplified by Carroll. From this viewpoint, downward causation seems mysterious because there is no clear mechanism for a whole to "push" its constituents. However, the problem dissolves if we adopt the difference-making framework, which more closely aligns with how scientists discover and exploit causal relations—one reason why the view has become so popular among philosophers of science. On this account, causation is not about physical production but about patterns of counterfactual dependence. The key question for downward causation becomes: if one were to *intervene* to change a higher-level variable, would the states of the lower-level variables reliably change as a result? Woodward argues that these variables represent real, objective properties of the system (2021, 191, n9), not merely pragmatic, coarse-grained descriptions. The upshot is that difference-making works equally well as an account of downward causation as it does for intra-level causation.

Consider feedback loops (Green and Batterman 2021, 55–56). A thermostat maintains room temperature through feedback control, operating with a setpoint—your desired temperature setting. When the system's sensors detect that the room temperature has moved from this setpoint, the thermostat responds by activating the heating or cooling system. Once the setpoint is reached, the device turns off. Notice that this higher-level setpoint setting directly influences lower-level physical processes: The setpoint value affects the motion of air molecules as they heat up and the flow of electrons through the system's circuitry. If you were to intervene and change the setpoint, this would reliably produce different patterns of molecular motion and electrical activity throughout the system. There is counterfactual dependence between these levels, the ability to manipulate the setpoint to bring about lower-level changes, and invariant behavior across many environments. This exemplifies downward difference-making causation. The system's goal-directed behavior is determined by higher-level properties rather than by an exhaustive understanding of its atomic components.

Interestingly, this same sort of feedback-and-response can be found in the regulation of blood sugar, body temperature, blood pressure, and a host of regularity systems in live organisms (Ellis 2016, 160). Biologists call it *homeostasis*. Negative feedback maintains the stability of a system subject to environmental changes. Again, the influence is top-down. Each system has an analogous setpoint (or range) for some higher-level property, which has a difference-making relation to the cellular level and below. Negative feedback and downward causation are not just found in humanly designed devices. They are essential to modern biology (Miller and Davidson 2013) (Ellis 2016, sec. 2.4.2).

Top-down causation in physics is often a matter of constraints placed on lower-level dynamics (Bishop, Silberstein, and Pexton 2022, 283–84; Green and Batterman 2021, 41–42). Some examples are familiar, such as the physical constraints on a violin that limit the way the strings can vibrate. Finger placements change these constraints. Likewise, a sail is constrained by its rigging. Different macro-level conditions influence the ways in which the constituent parts of the violin string and sail behave. A change in the upper-level boundary conditions brings about changes in the behavior of the constituents. Our ability to manipulate these constraints fits the difference-making account. Many of the "stability conditions" discussed in Section 7.4 work this way.

The point is that top-down causes are legitimate in terms of difference-making. That's good, given the popularity of difference-making among philosophers of science, for whom causation is a particularly important concept. Downward causation can be demonstrated through experimental interventions. There is nothing mysterious about it. "Top-down causation is not a suspicious, rare form of causation, but is ubiquitous in physical and biological systems alike" (Green and Batterman 2021, 59). Difference-making, including downward causes, is how causation is understood in MME.

9.2.3.4 Downward Causation and Determinism

There are consequences here for our understanding of determinism. Take a simple system like four particles in a closed container. That system can be represented by a phase space with twenty-four dimensions (three degrees of freedom for position and three for momentum for each particle). While the state of the system evolves continuously according to deterministic laws, introducing a fifth particle transforms the system itself—not merely its state. This intervention expands the phase space by adding six new dimensions, creating a different system. Both the four- and five-particle systems are deterministic, but the transition between them changes the framework within which deterministic laws operate.

This pattern appears in many physical systems. For example, a plucked violin string follows a deterministic equation of motion with a unique solution for any given finger placement (Sobolev 1964, 46–48).[10] Shifting the finger position alters the boundary conditions of that equation. The system before and after this adjustment is deterministic, but the change in boundary conditions is not. The finger placements constitute top-down interventions that shift the string from one deterministic configuration to another.

Note that these interventions don't violate physical laws—rather, the laws adapt to accommodate the new boundary conditions. This demonstrates how higher-level interventions can create discontinuities between otherwise deterministic systems without breaking the underlying physical principles. Left to itself, a system might be fully deterministic, but that does not bar the door to downward causation.

Difference-making downward causation is precisely what compatibilist libertarianism needs. Lower levels can be as deterministic as the vibrating string, but determinism at a given level is compatible with interventions from higher levels. This provides a way for understanding how mental causes can operate without violating physical laws: by setting boundary conditions that effectively select which deterministic pathway the system follows next.[11] That sort of selection is not all that downward causation does, however.

9.3 Opening Possibility Space

Jessica Wilson's weak metaphysical emergence (Section 7.5) involves the reduction of degrees of freedom in a system. She argues that lower levels have more freedom than emergent ones. An iron atom floating in space can travel along many different trajectories. That same atom contained in a steel plate is constrained. This elimination of degrees of freedom, she argues, renders the emergent entity ontologically distinct from its lower-level realizer, as fundamental physical laws cannot fully operate on or identify systems lacking all their original degrees of freedom.

There is an important difference between MME and Wilson's view. Downward causation and constraints not only reduce degrees of freedom—they can also *enable systems to reach possible states that were previously inaccessible*. This represents a fundamental insight about how emergence operates across multiple scales of reality. Consider some examples that illustrate this principle:

- Sailboats can move against the wind through a combination of macro-level aerodynamic and hydrodynamic principles. This technique,

called "tacking," involves zigzagging back and forth to move upwind. Changes in the macro-level constraints—specifically the rigging—make an action possible (moving upwind) that would otherwise be impossible (Green and Batterman 2021, 42).
- Lasers require highly engineered optical components and precise electrical inputs (Bishop 2024, 3-9–3-10). They produce highly coherent, single-wavelength beams of high intensity and focus by constraining photon pathways and allowed frequencies. The macroscopic constraints of the laser cavity bring about changes at the photon level, which in turn produces macroscopic effects in the observable beam.
- Quantum computing requires extreme isolation and low temperatures that enable quantum entanglement and superposition, which are difficult to maintain over significant periods. The macroscopic conditions induce quantum effects that can be harnessed for advanced computing applications.

Other examples range from mundane phenomena like pendulum dynamics (Bishop 2024, 3-9) and skeletal motion (Green and Batterman 2021, 51) to highly technical cases such as topological states in condensed matter physics (Qi and Zhang 2011). The lesson drawn from each example is the same: while constraints can limit behavior and reduce degrees of freedom as Wilson describes, they simultaneously enable novel behaviors by allowing systems to reach otherwise inaccessible states.

To understand how emergence enables free will, we must introduce the concept of "possibility space"—the range of potential states a system can access (Bishop 2024, sec. 3.5). Just as phase space represents all possible configurations of a physical system, possibility space encompasses all potentially realizable states given current constraints.[12] In many systems, certain regions of possibility space remain inaccessible under normal circumstances. These states are physically possible in principle, but they typically cannot be actualized without specific interventions. Emergence and downward causation can open areas of possibility space that were previously inaccessible, enabling systems to achieve capabilities they previously could not manifest.

Consider a specific example that demonstrates this principle clearly. A pennant displayed at my university was in orbit on the space shuttle Atlantis in May 2009. When that pennant was still in the factory, the state of being in orbit around Earth was completely inaccessible—there was simply no way for the pennant to achieve that state given its circumstances. That region of possibility space became actual due in part to the free choices made by NASA personnel. Without such decisions, those orbital states would have remained permanently inaccessible. This example illustrates how both the opening and closing of regions of possibility space are part of what libertarian free will accomplishes through downward causation. Free choices have tangible effects on bodies and their constituent atoms. Sometimes these choices restrict degrees of freedom, like the decision to place a cast on a broken arm. Sometimes they open areas of possibility space, like the decision to train for a marathon.

This opening and closing of possibility space regions operates in both directions. Emergence can decrease degrees of freedom, as Wilson demonstrates, but it can also produce novel causal powers at higher levels. Similarly, downward causation sometimes restricts the behavior of lower levels, but it also opens otherwise inaccessible regions of possibility space. The range of possible behaviors can increase or decrease in both upward and downward directions.

We can now state the conclusion of Chapter 8 in different terms. MME, through downward causation and the opening of possibility space, provides the metaphysical grounding for new, higher-level abilities. Free-will skeptics are correct about one important point: There is no free will at the level of fundamental physics. At the lowest levels of reality, the ability to do otherwise is inaccessible. However, emergence opens regions of possibility space. Through the complex interactions and constraints that characterize emergent systems, entirely new causal possibilities become available. Emergence opens new regions of possibility that would otherwise remain inaccessible.

The free-will skeptics in Chapter 1 are like the proverbial drunk looking for his car keys under the lamppost. Why? Because that's where the light is. For the question of free will, knowledge of fundamental physics is not sufficient. If emergence is real, there is more to reality at higher levels.

9.4 Compatibilist Libertarianism and MME

List's compatibilist libertarianism is a promising framework for understanding free will in light of modern physics. In this chapter, we considered List's three necessary conditions for free will: intentional action, alternative possibilities, and mental causation. The surprising element is his claim that these conditions can be satisfied even if physical determinism at the lowest levels is true.

While List provides the blueprint for compatibilist libertarianism, he acknowledges it remains programmatic without a detailed account of emergence. MME supplies precisely what his framework requires: a scientifically grounded explanation of how higher levels achieve genuine causal autonomy. For intentional action, MME counters reductionism by showing that higher levels in physics resist reduction to the micro level. The ubiquitous fundamentalist claim that all higher-level phenomena can be reduced "in principle" is less credible given that physics itself contains irreducible levels. For alternative possibilities, MME supports the reality of higher-level properties beyond mere coarse-graining, allowing for macro-indeterminism even with micro-determinism. For mental causation, the difference-making approach resolves Kim's causal exclusion problem by showing how mental states can make genuine causal differences at their own level while also influencing lower levels.

One last feature of MME is that emergence not only constrains lower-level dynamics but also opens regions of possibility space that would otherwise remain inaccessible. Free-will skeptics are correct that libertarian freedom is nowhere to be found at the level of fundamental physics. What they miss is how emergence opens the space of possibilities.

Compatibilist libertarianism provides an account of free will that neither requires substance dualism nor conflicts with physics. When bolstered by MME, it offers a promising research program for those who want to take physics and human agency seriously.

Notes

1 For some criticisms, see List et al. (2020).
2 Although the recent rise of panpsychism challenges this to some degree (Goff 2023). So, of course, does artificial intelligence.
3 This is a central idea in Husserl's account of intentionality. See Husserl (1982, sec. 84).
4 As I discussed in Chapter 2, leeway libertarianism posits that freedom requires the genuine ability to do otherwise, while sourcehood libertarianism insists that freedom requires the agent to be the ultimate originator of their actions, regardless of whether alternative possibilities exist.
5 The tyranny of scales problem describes how the mathematical representation for any given level of a system dictates a particular structure for matter at all lower scales, leading to inconsistencies between different levels of description.
6 In the examples we considered in Section 8.3, the causally relevant properties were at the meso level, including the number of spatial dimensions, temperature, and density.
7 "Partly" in the sense that it will also depend on the mechanical interaction of bones, tendons, and muscles.
8 The popular PhilPapers archive has 690 relevant papers, https://philpapers.org/browse/the-exclusion-problem.
9 See also the discussion on homeostasis in Section 9.2.3.3.
10 There are some idealizations in play here, such as a lossless string and end points that do not vibrate. I am also ignoring transients in the example that follow.
11 Downward causation also provides another reason for rejecting the causal closure of physics, as Ellis argues (2020).
12 In an actual mathematical state space, all the possible states of the system are of one type. For phase spaces, the states are typically given in terms of position

and momentum. Quantum mechanical systems require a different sort of state space. "Possibility space" lumps all these different sorts of systems together. Such a space cannot be rigorously defined.

Works Cited

Batterman, Robert W. 2021. *The Middle Way: A Non-Fundamental Approach to Many-Body Physics*. Oxford: Oxford University Press.

Bickle, John. 2020. "Multiple Realizability." In *Stanford Encyclopedia of Philosophy*, edited by Edward N. Zalta. Stanford: Metaphysics Research Lab, Stanford University. https://plato.stanford.edu/archives/sum2020/entries/multiple-realizability/.

Bishop, Robert C. 2024. *The Physics of Emergence*. 2nd ed. San Rafael: IOP Publishing. https://doi.org/10.1088/978-0-7503-6367-9.

Bishop, Robert C., and Peter Beim Graben. 2016. "Contextual Emergence of Deterministic and Stochastic Descriptions." In *From Chemistry to Consciousness*, edited by Harald Atmanspacher and Ulrich Müller-Herold. Cham: Springer International Publishing. https://doi.org/10.1007/978-3-319-43573-2_6.

Bishop, Robert C., Michael Silberstein, and Mark Pexton. 2022. *Emergence in Context: A Treatise in Twenty-First Century Natural Philosophy*. Oxford: Oxford University Press.

Carroll, Sean. 2016a. *The Big Picture: On the Origins of Life, Meaning, and the Universe Itself*. New York: Dutton.

Carroll, Sean. 2016b. "Consciousness and Downward Causation." *Preposterous Universe*, September 8. www.preposterousuniverse.com/blog/2016/09/08/consciousness-and-downward-causation/.

Caruso, Gregg D., Christian List, and Cory J. Clark. 2020. "Free Will: Real or Illusion – a Debate." *The Philosopher* 108 (1): 61–80.

Colyvan, Mark. 2024. "Indispensability Arguments in the Philosophy of Mathematics." In *Stanford Encyclopedia of Philosophy*, edited by Edward N. Zalta and Uri Nodelman. Stanford: Metaphysics Research Lab, Stanford University. https://plato.stanford.edu/archives/sum2024/entries/mathphil-indis/.

Dennett, Daniel. 1998. *The Intentional Stance*. Reprint. Cambridge: MIT Press.

Ellis, George F. R. 2016. *How Can Physics Underlie the Mind? Top-down Causation in the Human Context*. The Frontiers Collection. Berlin: Springer.

Ellis, George F. R. 2020. "The Causal Closure of Physics in Real World Contexts." *Foundations of Physics* 50 (10): 1057–97. https://doi.org/10.1007/s10701-020-00366-0.

Goff, Philip. 2023. *Why? The Purpose of the Universe*. Oxford: Oxford University Press.

Green, Sara, and Robert Batterman. 2021. "Making Sense of Top-Down Causation: Universality and Functional Equivalence in Physics and Biology." In *Top-down Causation and Emergence*, edited by Jan Voosholz and Markus Gabriel. Synthese Library 439. Berlin: Springer.

Hossenfelder, Sabine. 2019. "The Case for Strong Emergence." In *What Is Fundamental?*, edited by Anthony Aguirre, Brendan Foster, and Zeeya Merali.

The Frontiers Collection. Cham: Springer International Publishing. https://doi.org/10.1007/978-3-030-11301-8_9.

Husserl, Edmund. 1982. *General Introduction to a Pure Phenomenology*. Translated by F. Kersten. Husserliana: Edmund Husserl—Collected Works, Book 1. Dordrecht: Kluwer Academic.

Kahneman, Daniel, and Amos Tversky. 1979. "Prospect Theory: An Analysis of Decision under Risk." *Econometrica* 47 (2): 263. https://doi.org/10.2307/1914185.

Kim, Jaegwon. 2003. "Blocking Causal Drainage and Other Maintenance Chores with Mental Causation." *Philosophy and Phenomenological Research* 67 (1): 151–76.

Kim, Jaegwon. 2007. "Causation and Mental Causation." In *Contemporary Debates in Philosophy of Mind*, edited by Brian P. McLaughlin and Jonathan D. Cohen. Oxford: Blackwell.

List, Christian. 2019a. "What's Wrong with the Consequence Argument: In Defence of Compatibilist Libertarianism." *Proceedings of the Aristotelian Society* 119 (3): 253–74. https://doi.org/10.1093/arisoc/aoz018.

List, Christian. 2019b. *Why Free Will Is Real*. Cambridge: Harvard University Press.

List, Christian. 2022. "The Naturalistic Case for Free Will." In *Levels of Reality in Science and Philosophy*, edited by Stavros Ioannidis, Gal Vishne, Meir Hemmo, and Orly Shenker. Jerusalem Studies in Philosophy and History of Science. Cham: Springer International Publishing. https://doi.org/10.1007/978-3-030-99425-9_10.

Miller, Callie Johnson, and Lance A. Davidson. 2013. "The Interplay between Cell Signalling and Mechanics in Developmental Processes." *Nature Reviews Genetics* 14 (10): 733–44. https://doi.org/10.1038/nrg3513.

O'Connor, Timothy. 2000. *Persons and Causes: The Metaphysics of Free Will*. Oxford: Oxford University Press.

O'Connor, Timothy, and Christopher Franklin. 2021. "Free Will." In *Stanford Encyclopedia of Philosophy*, edited by Edward N. Zalta. Stanford: Metaphysics Research Lab, Stanford University. https://plato.stanford.edu/archives/spr2021/entries/freewill/.

Putnam, Hilary. 2012. "Indispensability Arguments in the Philosophy of Mathematics." In *Philosophy in an Age of Science: Physics, Mathematics and Skepticism*, edited by Mario De Caro and David Macarthur. Cambridge: Harvard University Press.

Qi, Xiao-Liang, and Shou-Cheng Zhang. 2011. "Topological Insulators and Superconductors." *Reviews of Modern Physics* 83 (4): 1057–110. https://doi.org/10.1103/RevModPhys.83.1057.

Quine, Willard V. 1948. "On What There Is." *The Review of Metaphysics* 2 (5): 21–38.

Sobolev, S.L. 1964. *Partial Differential Equations of Mathematical Physics*. 3rd ed. Edited by T. A. A. Broadbent. Translated by E. R. Dawson. New York: Dover Publications.

Timpe, Kevin. 2012. *Free Will: Sourcehood and Its Alternatives*. 2nd ed. New York: Continuum.

Wilson, Mark. 2017. *Physics Avoidance: Essays in Conceptual Strategy*. Oxford: Oxford University Press.
Woodward, James. 2003. *Making Things Happen: A Theory of Causal Explanation*. Oxford Studies in Philosophy of Science. Oxford: Oxford University Press.
Woodward, James. 2008. "Mental Causation and Neural Mechanisms." In *Being Reduced: New Essays on Reduction, Explanation, and Causation*, edited by Jakob Hohwy and Jesper Kallestrup. Oxford: Oxford University Press.
Woodward, James. 2021. "Downward Causation and Levels." In *Levels of Organization in the Biological Sciences*, edited by Daniel S. Brooks, James DiFrisco, and William C. Wimsatt. Cambridge: MIT Press. https://doi.org/10.7551/mitpress/12389.003.0013.

10 The Big Picture

Contemporary free-will skepticism rests on a fundamental conceptual confusion. Many physicists and philosophers have concluded that modern physics shows that libertarian free will is impossible. Throughout this book, I have argued that this conclusion conflates physics itself with philosophical interpretations of physical theories. The skeptics examined in Chapter 1—those who maintain that libertarianism is "incompatible with what we know about nature" (Hossenfelder 2022, 128) and requires violations of physical laws (Carroll 2016, 381)—have mistaken metaphysical commitments for scientific findings. Their arguments derive their force not from physics per se but from what I have called "fundamentalism": the philosophical thesis that bottom-up explanation is mandatory and that microphysical entities possess ontological primacy. This thesis, while prevalent among physicists, is neither entailed by physical theories nor philosophically required.

When scientists engage philosophical questions, they are not working from within their area of expertise. Why, then, use physicists as the primary foil for this book? There are plenty of philosophers who make better arguments for and against free will. Two reasons: first, the physicists that I've cited engage the questions, and they do so with an air of authority. That makes them fair game. Second, the public puts far more stock in the opinions of physicists than those of philosophers, regardless of how ill-informed those opinions happen to be. Physicists are simply not trained to address such issues as the foundations of ethics, the existence of God, or the possibility of free will. As the humanities continue to come under pressure in academia, philosophers cannot afford to ignore what scientists are saying.

For many physicists, the case against libertarianism boils down to the fact that we are made of atoms, atoms are governed by laws, and those laws make no allowances for free will. Many also affirm the existence of chairs, penguins, and all the things studied by the higher-level sciences.

DOI: 10.4324/9781003681199-10

But those things, we are told, are merely coarse-grained descriptions of their microphysical constituents. Penguins are real in the sense that they are recognizable, macroscopic patterns within the stuff that physics says exists. Because macroscopic objects are composed of real stuff—subatomic particles, quantum fields, etc.—they share in that reality. While the higher-level sciences are useful and perhaps indispensable tools for understanding these patterns, fundamentalists take the lowest physical level to be causally responsible for everything else. If we had a perfect understanding of the entities at the smallest scale and the computational capacity to solve the relevant equations, there would be no need for science other than particle physics.

My core argument has been that this picture stems not from physics itself but from fundamentalism. When fundamentalist philosophy is wedded to physics, the resulting hybrid is presented as though it were simply the dictates of science. The conflation is so pervasive that even sophisticated thinkers fail to recognize where physics ends and their philosophical commitments begin. To understand how libertarian freedom remains possible within a law-governed universe, we must carefully distinguish empirical discoveries from the philosophical framework in which they are typically embedded.

Many free-will skeptics believe that physics is deterministic at the level of our experience and perhaps at all lower levels (depending on one's interpretation of quantum mechanics). Given this belief, we began with the picture of physical determinism inherited from classical mechanics: forces acting on particles governed by a small number of laws. The Laplacian demon illustrates this picture. If classical mechanics were true, then a superintelligence could use it to precisely predict the future.

I argued in Chapter 3 that this portrayal is historically and scientifically overstated. Classical mechanics was never wholly deterministic. Determinism was more of an ideal for physics, one based on the prevailing metaphysics of the time.[1] Libertarians could argue that the case for physical determinism was never complete and is on even shakier ground given the discovery of quantum mechanics. I believe instead that libertarian freedom at the level of agents does not hinge on whether fundamental physics is deterministic or not. For the sake of argument, I granted that it is.

One way libertarians can escape the problem of determinism is to appeal to mind-body dualism. Even if the material world is deterministic, dualism says that a person is more than material and the immaterial part is not subject to the laws of nature. Of course, dualism itself faces objections from science. We considered two in Chapter 5: the causal closure of physics and the conservation of energy. Causal closure is the claim that for any physical event that has a cause, that cause is itself physical. For as often as causal closure is invoked in philosophy, its scientific credentials turn out to be rather thin. At some crucial stage of the argument, one often finds fundamentalist assumptions smuggled in. But dualists and other non-fundamentalists are under no obligation to adopt a rival metaphysics. As for conservation laws, Noether's theorem proves that they are conditional. If an immaterial mind is operating on a material system, conservation of energy does not apply to that system, and so there is no violation.

Granted, that's a big "if," which is why the book doesn't end in Chapter 5. Requiring dualism in order to believe in libertarian freedom is too high a price for many. But let me say to my dualist readers: go in peace. You don't have to hold an emergentist account of free will to see the value of the last few chapters. Dualists should be looking for robust alternatives to reductionism, not simply denying it because they believe in immaterial minds.

A different sort of determinism is presented by the block universe of special relativity. If the events that constitute our future are fixed and immutable, then there is no possibility of freely choosing to do otherwise. There is a fact of the matter, say, about what you are having for dinner exactly one month from today. In a block universe, that fact cannot be changed. It is part of the fabric of reality. As we saw, however, the block universe is not an unavoidable consequence of relativity. There are various ways to reconcile relativity with the open future needed by leeway libertarians, e.g., understanding it to be a theory of dynamical laws rather than the geometry of spacetime.

Even if the standard objections from determinism and the nature of time are not as conclusive as they initially seem, libertarians still have to contend with reductionism. As we saw in Chapter 6, most of the key examples in the reductionist narrative are incomplete or oversimplified. This includes the reduction of thermodynamics to statistical mechanics, which only applies to particular, idealized cases. There is no sense in which one level of physics has ever been fully reduced to another.

Today, philosophers and scientists have become more comfortable with the language of emergence. What they mean by that, however, is typically "weak emergence," which is merely an acknowledgment of our epistemic limitations. One can be a fundamentalist in terms of ontology and methodology but also concede that nature is too complex to see exactly how the levels fit together. Higher-level phenomena are "emergent" in the sense that we can't figure out how to reduce them. That's weak emergence.

Strong emergence, in contrast, is widely criticized for having little or no scientific support. Strongly emergent properties seem to pop into existence without any sort of continuity with their base. There doesn't seem to be any principled reason, for example, why consciousness is strongly emergent in humans but not in earthworms.

Fortunately, there is now conceptual space between weak and strong emergence. What I have called moderate metaphysical emergence (MME) is metaphysical in the sense that it is real and not merely a matter of our inability to reduce one level to another. While the best case for MME can be made in terms of biology,[2] our skeptics tend to dismiss arguments that are not grounded in physics. As we've seen, they accept higher-level sciences for pragmatic reasons but treat physics alone as realistic. This is why somewhat arcane topics like renormalization and multiscale modeling were needed to make the case for emergence. These practices reveal the causal autonomy of meso- and macro-scales and the irrelevance of micro-details for many higher-level phenomena. In short, physics supports MME.

Finally, MME strengthens the case for compatibilist libertarianism. In terms of List's three desiderata for free will, MME first undermines alleged "in principle" reductions of mental states and intentional agency. If such reductions aren't possible in physics, then promissory notes about the reducibility of higher-level phenomena seem far less plausible.

Second, List defends alternative possibilities (leeway libertarianism) by appealing to the reality of levels. Micro-determinism does not entail macro-determinism at the level of agents. MME supports this idea with a robust theory of emergence. There are physics-based reasons to believe in the autonomy of the macro from the micro. One cannot make blanket pronouncements about what is possible at the agential level from knowledge of the micro level.

Third, MME includes a difference-making account of top-down causation. This resolves Kim's exclusion argument and explains how mental states can causally influence physical states without violating physical laws. Mental causes are not sui generis.

In answer to physics-based, free-will skepticism, the claim that we are collections of particles governed by the laws of physics is true in a sense but is also profoundly misleading. Emergence shows that higher-level entities are not *just* collections of particles, and their causal capacities are not all reducible to fundamental laws. The Standard Model of particle physics is complete for its domain, but that domain is not the whole of reality. Downward causation from emergent levels does not add a new "force" to its equations, as Carroll suggests. It acts as a context or boundary condition that selects which deterministic pathway is actualized. Libertarian choice works with the laws of physics, not against them.

My approach has been interdisciplinary, bridging physics, philosophy of science, and philosophy of mind. It turns the tables on fundamentalism by using physics to defend free will and avoids both naive scientism and anti-scientific obscurantism. Moreover, the implications of emergence go beyond the question of libertarian freedom. Most scientists and philosophers are scientific realists about our mature theories. But how can one be a realist about, say, molecular biology if all that exists are subatomic particles and quantum fields? MME refutes the reductionist premise, allowing discoveries in biology to be treated in unapologetically realist terms. Even those who officially reject reductive physicalism, such as interactive dualists, too often tacitly accept its framework when it comes to material bodies. We need to recognize how deeply reductionism has shaped our intuitions and so misled us. In this, strong emergentists were right all along: We need to think differently about the world and reject the flattened ontology some tell us that science demands.

My prediction is that critics will simply move the goalposts. "Your MME isn't based on *fundamental* physics, so it doesn't count." That's fine. Let them continue to tell us what can be done "in principle." At some point, however, the promissory notes need to be paid off. I think we've waited long enough.

In the end, the conflict was never truly between philosophy and science (libertarian freedom and physics). It is a conflict between two philosophical views: libertarianism and a fundamentalist approach to physics. Physicalism and reductionism are fundamentalism's creeds, not conclusions drawn from science. Skeptics look for free will among the particles and fields of the Standard Model, and not finding it there, declare it nonexistent. But physics, seen in its entirety, does more than describe the ground floor of reality. It reveals the structure of an entire edifice, one with plenty of room for libertarian freedom.

Notes

1 This is reflected in Einstein's (now overused) assertion that God does not play dice. While indeterminism was not Einstein's primary criticism of quantum mechanics (Becker 2018, 51), it nonetheless fueled his doubts about the theory.

2 See, for example, Green (2021), Green and Batterman (2021), and Ellis (2025).

Works Cited

Becker, Adam. 2018. *What Is Real? The Unfinished Quest for the Meaning of Quantum Physics*. New York: Basic Books.

Carroll, Sean. 2016. *The Big Picture: On the Origins of Life, Meaning, and the Universe Itself*. New York: Dutton.

Ellis, George F. R. 2025. "Biological Emergence: A Key Exemplar of the Open Systems View." In *Open Systems: Physics, Metaphysics, and Methodology*, edited by Michael E. Cuffaro and Stephan Hartmann. Oxford: Oxford University Press.

Green, Sara. 2021. "Cancer beyond Genetics: On the Practical Implications of Downward Causation." In *Levels of Organization in the Biological Sciences*, edited by Daniel S. Brooks, James DiFrisco, and William C. Wimsatt, 195–214. Cambridge: MIT Press. https://doi.org/10.7551/mitpress/12389.003.0014.

Green, Sara, and Robert Batterman. 2021. "Making Sense of Top-Down Causation: Universality and Functional Equivalence in Physics and Biology." In *Top-down Causation and Emergence*, edited by Jan Voosholz and Markus Gabriel, 39–63. Synthese Library 439. Berlin: Springer.

Hossenfelder, Sabine. 2022. *Existential Physics: A Scientist's Guide to Life's Biggest Questions*. New York: Viking.

Index

Made in the USA
Las Vegas, NV
19 September 2026

94fb39e6-d0ce-4995-9004-ec172b3bb89cR01